Mike McGrath

Raspberry Pi 3

in easy steps

In easy steps is an imprint of In Easy Steps Limited
16 Hamilton Terrace · Holly Walk · Leamington Spa
Warwickshire · United Kingdom · CV32 4LY
www.ineasysteps.com

Notice of Liability
Every effort has been made to ensure that this book contains accurate
and current information. However, In Easy Steps Limited and the
author shall not be liable for any loss or damage suffered by readers
as a result of any information contained herein.

Trademarks
All trademarks are acknowledged as belonging to their respective
companies.

In Easy Steps Limited supports The Forest Stewardship Council (FSC),
the leading international forest certification organization. All our titles
that are printed on Greenpeace approved FSC certified paper carry the
FSC logo.

MIX
Paper from
responsible sources
FSC® C020837

Printed and bound in the United Kingdom

ISBN 978-1-84078-729-0

Contents

Preface

For such a small, inexpensive device the Raspberry Pi has become a huge global sensation. Creation of this book has provided me, Mike McGrath, a welcome opportunity to demonstrate features of the Raspberry Pi that encourage an interest in computing and programming. Example code listed in this book describes how to develop programs in easy steps – and the screenshots illustrate the actual results. I sincerely hope you enjoy discovering the exciting possibilities of Raspberry Pi and have as much fun with it as I did in writing this book.

In order to clarify the code listed in the steps given in each example, I have adopted certain colorization conventions. Components of the Python programming language are colored blue, programmer-specified names are red, numeric and string data values are black, and comments are green, like this:

```python
# Write the traditional greeting.
greeting = "Hello World!"
print( greeting )
```

Additionally, in order to identify each source code file described in the steps, a file icon and file name appears in the margin alongside the steps:

script.py   data.txt   image.gif   sound.wav   font.ttf

For convenience, I have placed source code files from the examples featured in this book into a single ZIP archive. You can obtain the complete archive by following these easy steps:

1 Browse to **www.ineasysteps.com** then navigate to Free Resources and choose the Downloads section

2 Find Raspberry Pi 3 in easy steps in the list, then click on the hyperlink entitled All Code Examples to download the archive

3 Next, extract the archive contents to your home directory folder, such as **/home/pi**

4 Now, call upon the Python interpreter from a Terminal to execute any code example, for instance issue the command **python hello.py** to see the output

1 Getting started

Welcome to the exciting world of the Raspberry Pi. This chapter demonstrates how to establish a fully functional computer system.

Introducing Raspberry Pi

The Raspberry Pi is an inexpensive computer built on a single printed-circuit board. It was developed in the UK by the Raspberry Pi Foundation to encourage the teaching of basic computer science in schools and to put the fun back into learning about computing. The foundation recognized that the school ICT curriculum had changed, placing emphasis on the use of applications, such as Word and Excel, or to writing web pages. Additionally, they noticed that the home PC and games console had replaced the Amigas, BBC Micros, Spectrum ZX and Commodore 64 machines that people of an earlier generation learned to program on. Nowadays, young people have become merely passive users of computers who the foundation considers could benefit from knowing how computers work and how to program them – so they created the cheap, accessible Raspberry Pi computer.

Discover more about Raspberry Pi online at **raspberrypi.org**

In order to keep the price low, the Raspberry Pi 3 Model B has some innovative design features:

- At its heart is an ARM processor that has System-on-Chip (SoC) architecture to integrate several traditionally separate components onto a single chip. The ARM processor runs at 1.2GHz. Typically, ARM processors have previously been used mainly in cellphones.

- Unlike traditional computer design, the Raspberry Pi does not have a hard drive but instead employs a Micro SD card to contain the operating system and to store the files you create. The operating system can be one of several specially optimized variants of the Linux operating system or Windows 10 IoT.

The choice of Linux distributions optimized for the Raspberry Pi are described on page 12.

- The Raspberry Pi 3 Model B has a total memory of just 1GB – which is small compared to that of today's traditional computers. Even with this limitation, surprisingly good performance is achieved as neither the processor nor the operating system are "memory hungry". This in turn allows programs running on the Raspberry Pi to use very low amounts of memory.

- Most noticeably, each Raspberry Pi is supplied without a case so it can be easily built into another device, such as a monitor, and its components can be easily identified.

Raspberry Pi 3 Model B
Size: 85.6mm x 56mm x 21mm.

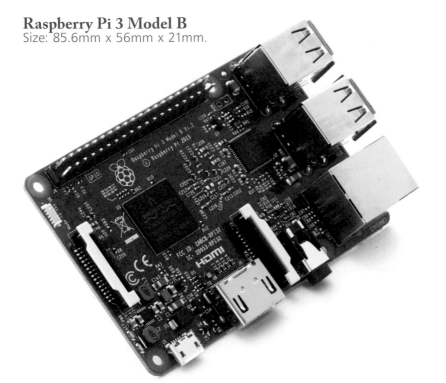

The Raspberry Pi 3 has an identical form factor to the previous Pi 2 (and Pi 1 Model B+) and has complete compatibility with Raspberry Pi 1 & 2.

Component:	Specification:
CPU	1.2GHz 64-bit Quad-core ARMv8
SoC	Broadcom BCM2837 chipset
GPU	Dual-core VideoCore 4 3D
RAM	1GB LPDDR2
Network	802.11 b/g/n Wireless LAN 10/100 Ethernet port (RJ45) Bluetooth 4.1 (Classic & Low Energy)
USB	4 x USB 2.0 ports
GPIO	40 header pins
Video	Full HDMI port
Audio	3.5mm jack and composite video
Camera	Camera Serial Interface (CSI)
Display	Display Serial Interface (DSI)
Storage	Micro SD card slot (push-pull)

The Raspberry Pi 3 Model B is recommended for use in schools and for general use, but the Pi Zero and the Pi 1 Model A+ are useful for embedded projects that require low power.

Gathering the components

The first step in establishing a computer system with Raspberry Pi is to gather together all the necessary components listed below:

- **Raspberry Pi** – first you will need to get your hands on a Raspberry Pi board itself, available worldwide through Premier Farnell/Element 14, Allied Electronics, and RS Components.

- **Micro SD card** – to contain the operating system, so is recommended at least 8GB capacity and Class 4 or higher.

- **Micro SD card reader** – to write the operating system onto the card, unless you have pre-installed system card.

- **Micro USB 5 volt power source** – typically an cellphone charger or e-book reader charger, providing a power supply output of at least 5V.

- **USB mouse** – any standard mouse.

- **USB keyboard** – any standard keyboard.

- **TV or monitor** – any HDMI/DVI monitor or TV should work, but for best results use one with HDMI input.

- **HDMI cable** – to connect to a TV/monitor.

- **Ethernet cable** – for wired internet connection, unless you use the built-in Wi-Fi adapter to connent to a router wirelessly.

This seemingly long list of requirements is broadly similar to the components of a traditional PC so you may well already have some items on this list, but others you may need to purchase.

You can find the Raspberry Pi suppliers online at **farnell.com** and at **alliedelec.com** or at **rs-components.com**

Don't confuse the larger Mini USB standard socket, found on many digital cameras, with the Micro USB standard power socket found on the Raspberry Pi.

Raspberry Pi 3 Model B

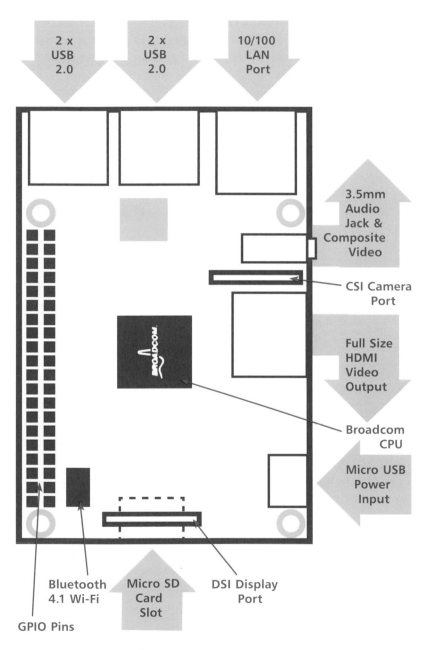

2 x USB 2.0

2 x USB 2.0

10/100 LAN Port

3.5mm Audio Jack & Composite Video

CSI Camera Port

Full Size HDMI Video Output

Broadcom CPU

Micro USB Power Input

Bluetooth 4.1 Wi-Fi

Micro SD Card Slot

DSI Display Port

GPIO Pins

The Camera Serial Interface (CSI) port allows connection of the official Camera Module accessory – an extra available from Raspberry Pi stockists.

The Display Serial Interface (DSI) port allows connection of the official 7" Touchscreen Display accessory – an extra available from Raspberry Pi stockists.

When you have gathered together all necessary components they can be connected as shown above – but the Raspberry Pi is not yet functional as the SD card doesn't contain an operating system.

Setting up the SD card

The operating system used by the Raspberry Pi can be one of several specially optimized versions of the Linux operating system. These are known as "distributions", or "distros" for short. Each available distro is offered as a disk image which must be written onto the Micro SD card to be inserted into the Raspberry Pi. The choice of currently available distros are listed for free download at **raspberrypi.org/downloads**
At the time of writing, the list includes these distros:

Raspbian is a free operating system based on Debian Linux and optimized for Raspberry Pi hardware.

- **Raspbian** – recommended for beginners, this distro provides a comprehensive system with a graphical desktop, web browser, and development tools to get you started programming.
- **Ubuntu MATE** – regular Ubuntu desktop distro optimized for Raspberry Pi 2 and Raspberry Pi 3.
- **OSMC** – Open Source Media Center that allows media playback from your local network, attached storage, or the web.
- **OpenELEC** – embedded operating system that provides an open source media hub.
- **PiNet** – classroom operating system that provides centralized student accounts and file storage.
- **RISC OS** – non-Linux alternative operating system.
- **Windows10 IoT** – Microsoft system for Core devices.

Formatting the SD card
It is recommended that you format the Micro SD card before writing any distro onto the card to ensure best performance:

The Micro SD card should be at least 8GB capacity and at least Class 4 speed.

1 Visit **sdcard.org/downloads/formatter_4** to download a free SD Formatter for Windows or Mac

2 Follow the instructions to install the SD Formatter, then insert your Micro SD card into your card reader and make a note of the drive letter allocated to it, e.g. **E:/**

3 Run SD Formatter then select the drive letter of your Micro SD card and format the card

Installing the Raspbian operating system

This book demonstrates the officially recommended Raspbian distro that can be written onto your Micro SD card using the "NOOBS" (**N**ew **O**ut **O**f the **B**ox **S**oftware) easy installer.

1 Go to **raspberrypi. org/downloads/ noobs** and click the **Download ZIP** button to download the installer

You can alternatively purchase a pre-installed NOOBS Micro SD card from Raspberry Pi retailers.

2 Extract all contents in the downloaded ZIP file to your Micro SD card to copy the installer

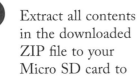

The drive letter assigned to the SD card on your system may be different to the one shown here.

3 Remove the Micro SD card from the card reader and insert it into your Raspberry Pi

4 Plug in your keyboard, mouse and monitor cables

5 Now, plug in the USB power cable to your Raspberry Pi to see it boot to a window offering a list of operating systems you can install

6 Check the box next to the "Raspbian" option, then click the **Install** button to install the operating system

7 When the installation process has completed, you can turn the power off, then back on, to start the operating system

Configuring the system

After you start up your Raspberry Pi you can set up various system options using the configuration tool:

 Click the "Terminal" launcher button to open a Terminal window that provides a command line prompt

Terminal

 Precisely type the command **sudo raspi-config** at the prompt, then hit Return to open the configuration tool

The **sudo** command provides administration privileges that allow system configuration changes.

If the Raspberry Pi screen does not fill your monitor, choose the Advanced Options' "Overscan" configuration menu item then select the "Disable" option.

---

pi@raspberrypi: ~

File Edit Tabs Help

pi@raspberrypi:~ $ sudo raspi-config

---

pi@raspberrypi: ~

File Edit Tabs Help

```
┤ Raspberry Pi Software Configuration Tool (raspi-config) ├

    1 Expand Filesystem          Ensures that all of the SD card s
    2 Change User Password       Change password for the default u
    3 Boot Options               Choose whether to boot into a des
    4 Wait for Network at Boot   Choose whether to wait for networ
    5 Internationalisation Options  Set up language and regional sett
    6 Enable Camera              Enable this Pi to work with the R
    7 Add to Rastrack            Add this Pi to the online Raspber
    8 Overclock                  Configure overclocking for your P
    9 Advanced Options           Configure advanced settings
    0 About raspi-config         Information about this configurat

              <Select>                        <Finish>
```

 Use the keyboard up/down arrow keys to highlight an item you wish to configure

 Hit the Return key to select a highlighted item and reveal its options, e.g. choose "Internationalisation Options"

5 Use the arrow keys to now choose "Change Locale" and deselect the current character set from the list then select a character set for your locale – such as "en_US.UTF-8"

```
┤ Raspberry Pi Software Configuration Tool (raspi-config) ├
I1 Change Locale          Set up language and regional sett
I2 Change Timezone        Set up timezone to match your loc
I3 Change Keyboard Layout Set the keyboard layout to match
I4 Change Wi-fi Country   Set the legal channels used in yo




            <Select>                    <Back>
```

6 Choose "Change Timezone" and select your region and timezone – such as "America" then "Eastern"

```
┤ Configuring tzdata ├
Please select the city or region corresponding to your time zone.

Time zone:
                    Alaska
                    Aleutian
                    Arizona
                    Central
                    Eastern
                    Hawaii
                    Starke County (Indiana)
                    Michigan
                    Mountain
                    Pacific Ocean
                    Pacific-New
                    Samoa

            <Ok>                    <Cancel>
```

You can re-visit the configuration menu at any time from a command prompt by issuing the command **sudo raspi-config**.

7 Press the right arrow key to select "OK", then hit Return to apply the configuration change

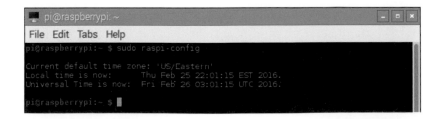

```
pi@raspberrypi: ~
File Edit Tabs Help
pi@raspberrypi:~ $ sudo raspi-config

Current default time zone: 'US/Eastern'
Local time is now:      Thu Feb 25 22:01:15 EST 2016.
Universal Time is now:  Fri Feb 26 03:01:15 UTC 2016.

pi@raspberrypi:~ $
```

Logging into the console

Where the Raspbian operating system is configured to not automatically start the graphical desktop, your Raspberry Pi will start up with the boot messages then display a login prompt where you must enter a username and password to proceed.

The Raspbian system is supplied configured with a single default user named "pi" whose login password is "raspberry". These can therefore be used at the login prompt to access the system at its Command-Line Interface (CLI). This is known as the "console" – historically describing the keyboard and monitor combination from the days before the mouse and graphical desktop. When a login attempt succeeds, a colorful user command prompt appears where you can enter commands to interact with the system.

If you make a mistake entering the username or password you will simply be returned to the login prompt so you can try again.

 Start up your Raspberry Pi and watch the boot process until a login prompt appears – type **pi** then hit Return

 When the password prompt appears, type **raspberry** then hit Return – see the user prompt appear after system info

Linux is case-sensitive so you must always observe correct capitalization. For example, **pi**, **Pi**, and **PI** are distinctly different.

```
raspberrypi login: pi
Password:
Last login: Sun Mar 27 15:36:22 EEST 2016 on tty1
Linux raspberrypi 4.1.18-v7+ #846 SMP Thu Feb 25 14:22:53 GMT 2016 armv7l

The programs included with the Debian GNU/Linux system are free software;
the exact distribution terms for each program are described in the
individual files in /usr/share/doc/*/copyright.

Debian GNU/Linux comes with ABSOLUTELY NO WARRANTY, to the extent
permitted by applicable law.
pi@raspberrypi:~ $ 
```

The Raspberry Pi user prompt contains several components describing your username **pi**, machine name **raspberrypi** and current file location ~ (tilde – an alias for the **/home/pi** directory). You can confirm your current location on the filesystem at any time by entering the command **pwd** (print working directory).

 At the user prompt, type **pwd** then hit Return – see the current directory appear then another user prompt

The password will not be displayed as you type it at the password prompt – so it cannot be stolen from over your shoulder.

```
pi@raspberrypi ~ $ pwd
/home/pi
pi@raspberrypi ~ $ 
```

You can discover the contents of the current directory by entering the command **ls** (list sorted). This displays a list of folders and files within the current directory, but not hidden system files. Like most Linux commands, "switches" can be added after the command to change how they respond. So with the **ls** command, adding a **-a** switch will display all content, including system files.

  At the user prompt, type **ls** then hit Return – see the current directory contents appear

```
pi@raspberrypi ~ $ ls
Desktop  python_games
pi@raspberrypi ~ $
```

  Now, at the user prompt, type **ls -a** then hit Return – see the current directory contents including system files

```
pi@raspberrypi ~ $ ls -a
.              .bashrc   Desktop  python_games
..             .cache    .gvfs    .thumbnails
.bash_history  .config   .local   .Xauthority
.bash_logout   .dbus     .profile .xsession-errors
pi@raspberrypi ~ $
```

You can clear past command results using the **clear** command and discover switches available for any command, plus what they achieve, using the **man** command followed by the command name, e.g. **man ls** to display the switches for the **ls** command.

It is important to recognize that Linux is designed as a multi-user system, so regular users cannot perform all tasks – some can only be performed by the system administrator ("root" or "superuser"). But for convenience you can temporarily assume superuser status to do something with the **sudo** command, for example to close Raspberry Pi with the **shutdown** command. This accepts a **-h** (halt) switch or a **-r** (reboot) switch, followed by a number indicating a delay in minutes, or zero for no delay.

  Finally, at the user prompt, type **sudo shutdown -h 0** then hit Return – see the system close down immediately

```
pi@raspberrypi ~ $ sudo shutdown -h 0
```

Switches can be combined, so that **ls -al** is equivalent to **ls -a -l** as both will display all content in long format.

Notice that system content names are prefixed by a period, and directories are blue color.

The system remembers your past commands – use the up and down arrow keyboard keys to scroll through them at the user prompt then hit Return to re-run your chosen past command.

You can use Boot Options in the configuration tool to change how your system starts up (see page 20).

Starting the desktop

Interacting with your Raspbian system at the Command-Line Interface (CLI) gives an insight into the power of commands. Many Linux "boxes" are used at just that level, especially web servers, but you will probably want to see a more familiar graphical desktop that lets you interact using your mouse.

With the Raspbian system, an "X" server provides a Graphical User Interface (GUI) that automatically employs a window manager to control running application windows.

When you boot Raspbian, the X server will not automatically get started if you configure your system as a basic console. Simply issuing a **startx** command at the command prompt will start the X server and load the graphical desktop environment.

1 Start up your Raspberry Pi then log in to the system as the default user **pi** with the **raspberry** password

2 When the user prompt appears, type **startx** then hit Return – see the graphical desktop load in the screen

```
pi@raspberrypi ~ $ startx
```

...cont'd

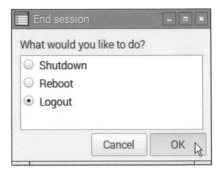

3 Click the Menu button then select the Shutdown menu item – to launch the "End session" dialog box

Features of the graphical desktop are described in the next chapter – but you may like to explore the menu items now.

4 Choose "Logout" then click **OK** to close the desktop

5 Enter your username and password to restart the desktop

If you wish to shutdown the system you can do so from the "End session" dialog or from a command prompt in a Terminal window or in a basic console.

1 Click the "Terminal" launcher button to open a Terminal window that provides a command line prompt

You can restart the system using the Reboot option in the "End session" dialog or from a command prompt by issuing the command **sudo shutdown -r 0**.

2 Precisely type the command **sudo shutdown -h 0** at the prompt, then hit Return to completely close the system

Automating the login

Unless you particularly wish to use your Raspberry Pi as a console system it becomes tedious to log in and start the X server manually, so the process can be automated using the configuration tool.

 Start up your Raspberry Pi then log in to the system as the default user **pi** with the **raspberry** password

 When the user prompt appears, type **sudo raspi-config** then hit Return – see the configuration tool appear

pi@raspberrypi ~ $ sudo raspi-config

 When the configuration menu appears, use the up/down keyboard arrow keys to choose the "Boot Options" item – then hit Return to select it

The "Overclock" menu item allows the ARM processor on some models of the Raspberry Pi to be run above their default speed to increase performance – but this may reduce the life of your Raspberry Pi.

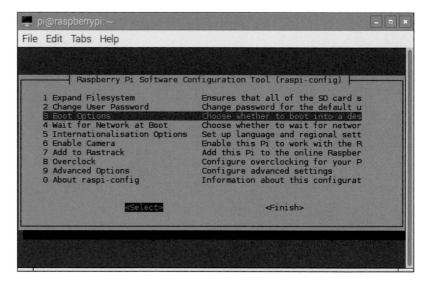

pi@raspberrypi: ~

File Edit Tabs Help

```
┤ Raspberry Pi Software Configuration Tool (raspi-config) ├

1 Expand Filesystem            Ensures that all of the SD card s
2 Change User Password         Change password for the default u
3 Boot Options                 Choose whether to boot into a des
4 Wait for Network at Boot     Choose whether to wait for networ
5 Internationalisation Options Set up language and regional sett
6 Enable Camera                Enable this Pi to work with the R
7 Add to Rastrack              Add this Pi to the online Raspber
8 Overclock                    Configure overclocking for your P
9 Advanced Options             Configure advanced settings
0 About raspi-config           Information about this configurat

              <Select>                      <Finish>
```

 In the list of options that appears next, use the up/down keyboard arrow keys to choose the "Desktop Autologin" item – then hit Return to select it

```
┤ Raspberry Pi Software Configuration Tool (raspi-config) ├
  B1 Console           Text console, requiring user to login
  B2 Console Autologin Text console, automatically logged in as 'pi' user
  B3 Desktop           Desktop GUI, requiring user to login
  B4 Desktop Autologin Desktop GUI, automatically logged in as 'pi' user




                <Ok>                      <Cancel>
```

5 Next, use the left-right keyboard arrow keys to choose "Finish" then hit Return to apply the configuration change

```
┤ Raspberry Pi Software Configuration Tool (raspi-config) ├
  1 Expand Filesystem          Ensures that all of the SD card s
  2 Change User Password       Change password for the default u
  3 Boot Options               Choose whether to boot into a des
  4 Wait for Network at Boot   Choose whether to wait for networ
  5 Internationalisation Options Set up language and regional sett
  6 Enable Camera              Enable this Pi to work with the R
  7 Add to Rastrack            Add this Pi to the online Raspber
  8 Overclock                  Configure overclocking for your P
  9 Advanced Options           Configure advanced settings
  0 About raspi-config         Information about this configurat

            <Select>                        <Finish>
```

Avoid changing the user password "raspberry" for the default user named "pi" in case you forget the new password.

21

6 In the dialog that now appears use the left/right keyboard arrow keys to choose "Yes" – then hit Return to reboot the system straight to the GUI desktop

```
  Would you like to reboot now?







            <Yes>                  <No>
```

Summary

- Raspberry Pi is an inexpensive credit-card sized computer developed to put fun back into learning about computing.

- Raspberry Pi has an ARM processor using System-on-Chip architecture and uses an SD card as its hard drive for economy.

- Peripheral components must be connected to the Raspberry Pi board to establish a complete computer system.

- The Raspbian Linux distro is recommended for beginners and must be written onto an SD card for use with Raspberry Pi.

- The Raspbian system has a configuration tool **raspi-config** in which you can change system options.

- The Raspbian distro is supplied configured with a default user name of "pi" whose login password is "raspberry".

- In a basic console, a login prompt appears in the Command-Line-Interface (CLI) after the boot messages have completed.

- After login, you can confirm your location on the filesystem by entering the command **pwd** (print working directory).

- You can discover the contents of the current directory folder by entering the command **ls** (list sorted).

- The **clear** command removes past command results and the **man** command can be used to discover command switches.

- Linux systems, such as Raspbian, are designed as multi-user systems so regular users cannot perform administrative tasks.

- You can temporarily assume superuser status to perform administrative tasks with the **sudo** command.

- With Raspbian, the X server provides a Graphical User Interface (GUI) desktop with a window manager.

- The desktop GUI can be started from a console prompt by entering the command **startx** or it can be automated by modifying the "Boot Options" with the configuration tool.

2 Exploring the desktop

This chapter demonstrates features of the Raspbian GUI desktop environment.

Understanding the taskbar

Usefully Raspbian provides a default taskbar containing an **Application Launcher,** which provides buttons to launch the most frequently used applications, and a **System Tray** that provides frequently sought system information.

The System Tray, contains these five default items:

Wi-Fi Volume CPU Clock Ejecter
Monitor

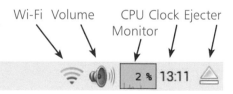

All these System Tray items, except the CPU Monitor, are buttons that you can click to adjust their settings.

- **Wi-Fi Network Settings** – displays available networks and lets you connect to your selected network
- **Volume control** – displays the current volume level and allows you to adjust the volume
- **CPU Usage Monitor Settings** – displays current CPU load
- **System clock** – displays the current system time and date
- **Ejecter button** – allows you to safely eject external drives that may be connected to your Raspberry Pi

...cont'd

The Application Launcher buttons are used to launch the application **Menu** ("Start" button), **Epiphany Web Browser**, **File Manager**, and the **Terminal** command window. The taskbar also provides an area to contain buttons for minimized windows.

Menu   Web Browser   File Manager   Terminal   Minimized Window area

Your Raspbian operating system may provide more buttons in the Application Launcher panel in addition to the core default buttons illustrated here.

- **Menu** – pops open a comprehensive application menu
- **Web Browser** – launches the Epiphany web browser
- **File Manager** – launches a graphical file management application in your home directory, such as **/home/pi**
- **Terminal** – launches a console-like command window

Follow these steps to see the initial content within your home directory, where all the files you create will be stored:

 Click the File Manager button on the taskbar – to launch the File Manager application window

 Choose **Folder View Mode, Icon View** in the File Manager's **View** menu – to see your home content

Choose **View**, **Show Hidden** in the File Manager to reveal hidden system files and folders. These have names that begin with a period and should not be modified or deleted. Deselect **View**, **Show Hidden** to ensure these remain hidden from view in the future.

File Manager window

```
pi
File  Edit  View  Bookmarks  Go  Tools  Help
          /home/pi
Directory Tree
 pi
   Desktop          Desktop  Documents  Downloads  Music  Pictures  Public
   Documents
   Downloads
   Music            Templates  Videos
   Pictures
   Public
   Templates
   Videos
 /
8 items (18 hidden)                    Free space: 2.1 GiB (Total: 5.8 GiB)
```

Choosing your preferences

If the default appearance of the Raspbian GUI desktop is not to your taste you can easily change the look of the window widgets, colors, icons, cursors, and fonts by choosing alternative preferences.

 Click **Menu**, **Preferences**, **Appearance Settings** – to launch the Appearance Settings dialog

 Use the options to select your preferences for **Desktop** colors, layout, and wallpaper

 Select options for the **Menu Bar** (taskbar) size, position, and colors

 Choose font and color preferences for overall **System** appearance

Your selections are applied temporarily to provide a preview – click the **OK** button to apply them permanently

Use the system highlight color to change the window menu bar color, shown here in red.

The appearance of the desktop itself can also be easily changed by choosing alternative preferences for the wallpaper, background color, and text label font, color and shadow.

6 Right-click anywhere on the desktop then choose **Desktop Preferences** from the context menu that appears – to launch the Desktop Preferences dialog

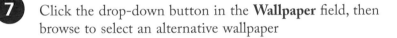

Create New...   ›

🗎 Paste

🗐 Select All
     Invert Selection

Sort Files   ›

🖺 Desktop Preferences ▶

| Appearance | Desktop Icons | Advanced |

Background

Wallpaper mode: Center unscaled image on the monitor   ▾

Wallpaper: ▣ paint-splash.jpg   ▣

☑ Use the same wallpaper on all desktops

Background colour: ▢   ˉ

Text

Label text Font: Roboto Light   12

Label text Colour: ▇   Shadow Colour: ▢

You can browse the internet for suitable wallpaper images and use the Wallpaper mode options to fit the image to your monitor screen.

7 Click the drop-down button in the **Wallpaper** field, then browse to select an alternative wallpaper

8 Choose alternative preferences for the other options if desired, then click the **Close** button to apply your choices to the desktop

27

Examining the filesystem

If you are unaccustomed to a Linux operating system, the Raspbian filesystem may seem unusual. It is arranged in a hierarchical tree with a / location at its top "root" location.

Beneath the / root location are 20 named directories (folders) that mostly contain system files. For example, the "bin" directory that has the filesystem location address of **/bin** – combining the root location of **/** and its directory name.

The **/home** directory is most worthy of note as it contains a sub-directory for each named user registered on the system. Files created by each user are stored within the directory bearing their username, or sub-directories within that username directory. For example, the "pi" directory for the default user has the filesystem location address of **/home/pi**, where documents might be stored by that user in the sub-directory at **/home/pi/Documents**.

The Raspbian filesystem can be examined using the File Manager to better understand its hierarchical structure:

1 Click **Menu, Accessories, File Manager,** or click the **File Manager** button on the taskbar – to launch the File Manager tool

Do not confuse the "root" location at **/** with the "root" superuser (who has a root folder).

In the **Places** pane, click on **Desktop** to see desktop shortcut icons, and **Applications** to see all Menu categories.

By default, the File Manager opens in the current user's home directory. Notice that the File Manager window's title bar displays the name of the directory whose contents it displays, and the filesystem address of that directory is shown in its address field. The Places pane also highlights the Home Folder location.

...cont'd

2 Next, click the arrow button on the File Manager toolbar to move up to the next hierarchy level

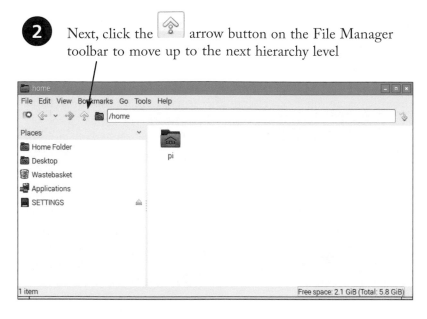

In the **Places** pane, click on **Wastebasket** to reveal all deleted files. Right-click on any deleted file, then choose **Restore** on the context menu that appears – to return that file to its original filesystem location.

The File Manager window's title bar now displays the name of the "parent" directory containing your home directory, and the filesystem address of that directory is shown in its address field.

3 Now, click the button once more to move up to the next hierarchy level – to see all "child" directories contained within the top-level **/** root directory

bin	boot	dev	etc	home	lib
lost+found	media	mnt	opt	proc	root
run	sbin	srv	sys	tmp	usr
var					

19 items   Free space: 2.1 GiB (Total: 5.8 GiB)

Never edit content within any directory other than your own home directory (or sub-directories therein) unless you know exactly what you are doing – or you may corrupt the system configuration.

You can use the same commands described for the console, on page 17, in a terminal window – for example use **ls** to list directory contents.

Launching a terminal

With the GUI desktop automatically starting when Raspbian boots up, a "Terminal" window provides a prompt that allows commands to be executed as if at the console CLI prompt. Filesystem navigation can be easily performed at a command prompt by issuing a **cd** (change directory) command followed by a location address. For example, **cd /home** navigates to the "home" directory. Shorthand commands of **cd ..** navigates up one hierarchy level and **cd ~** navigates to the current user's home directory from anywhere.

 Click **Menu**, **Accessories**, **Terminal**, or click the Terminal button on the taskbar – to launch a terminal window

The appearance of the terminal window can be easily changed. Click **Edit**, **Preferences** on the terminal window menu and choose new Background and Foreground colors.

At the terminal prompt, enter the command **pwd** then hit enter to see the current location on the filesystem

```
pi@raspberrypi: ~
File  Edit  Tabs  Help
pi@raspberrypi:~ $ pwd
/home/pi
pi@raspberrypi:~ $
```

By default, the terminal opens in the current user's username directory, and its title bar and the prompt both display the shorthand name of the directory as **~**.

...cont'd

3 Enter the **cd ..** command to navigate up the filesystem hierarchy, and the **pwd** command to confirm location

```
pi@raspberrypi:~ $ pwd
/home/pi
pi@raspberrypi:~ $ cd ..
pi@raspberrypi:/home $ pwd
/home
pi@raspberrypi:/home $ cd ..
pi@raspberrypi:/ $ pwd
/
pi@raspberrypi:/ $ 
```

Notice that both the terminal window's title bar and the prompt display the current directory location after each move.

4 Enter the **clear** command to remove the previous results from the terminal window, then enter the **cd ~** command to return to the username directory in a single move

```
pi@raspberrypi:/ $ cd ~
pi@raspberrypi:~ $ pwd
/home/pi
pi@raspberrypi:~ $ 
```

Some of the system directories contain useful binary executable files that can be addressed from the command prompt only. For example, the **/opt/vc/bin/tvservice** executable file is useful when your Raspberry Pi is connected to a TV by HDMI cable.

5 Enter the command **/opt/vc/bin/tvservice -s** to discover the screen resolution of a TV connected by HDMI

```
pi@raspberrypi:~ $ /opt/vc/bin/tvservice -s
state 0x12000a [HDMI CEA (4) RGB lim 16:9], 1280x720 @ 60.00Hz, progressive
pi@raspberrypi:~ $ 
```

Some commands can only be performed by the superuser – but you can temporarily assume superuser status with the **sudo** command.

Executable file switch options can typically be discovered with a **-h** or a **--help** switch. Here, the **-s** switch displays a status revealing a screen resolution of 1280x720.

31

Creating a text file

Raspbian provides a plain text editor named Leafpad that works much like the Notepad plain text editor in Windows. Text files can be simply created with Leafpad then saved in your username directory – or a sub-directory within that directory:

Plain text lines in Linux end with an invisible line feed character (LF) but on Windows they end with both carriage return and line feed characters (CR + LF). So plain text files created on Linux appear to have no line endings when viewed on Windows unless the endings are converted.

1 Click on **Menu, Accessories, Text Editor** to launch the Leafpad plain text editor

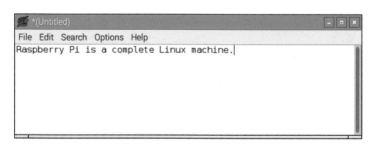

2 Next, type some text in the Leafpad editor window

The Options menu in Leafpad lets you choose Font, Word Wrap, and Line Number preferences.

3 Now, click the **File** item on the Leafpad menu, then choose the **Save As** option – to launch the Save As dialog

By default, the Save As dialog will automatically offer to save files in your home directory but you can choose an alternative sub-directory such as "Documents" if desired.

 Type a file name of your choice, and optionally a file extension, into the **Name** field

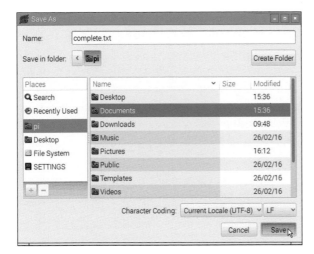

You can quickly create a new sub-directory by clicking the **Create Folder** button in the Save As dialog and typing a name.

5 Finally, click the **Save** button to create the text file with your given name in your chosen location

In Linux, filename extensions, such as **.txt** are not an essential requirement of the system but help to readily identify the file type to the user.

Plain text files in Raspbian are associated with Leafpad by default.

6 Close Leafpad then double-click on the file icon of the plain text file you just created – to see Leafpad launch once more, displaying that text file for editing

By default, the Raspbian boot process assigns an IP address to the ethernet socket device (eth0) using Dynamic Host Configuration Protocol (DHCP) – so manual configuration should not be required.

To connect to your router wirelessly, simply click the Wi-Fi button on the System Tray, then choose the network and enter its security key code (password).

Browsing the internet

Connecting to the internet first requires a network cable from your router to be plugged into the Raspberry Pi's ethernet socket, or a Wi-Fi connection to be established with the router. When the Raspbian operating system boots up, it then automatically detects the connection and configures an IP address for the socket so you are instantly online. To confirm that the internet is connected, you can issue a **ping** command at a terminal prompt to see a response from a specified network location host – either as a numerical IP address, such as **192.168.0.1**, or as a named URL address such as **google.com**

The **ping** command attempts to send a tiny "packet" of data to the specified location. When the attempt succeeds, the host returns a response to your system so you know the internet is connected. This will continue unless the command specifies a count value with a **-c** switch, e.g. **-c 3** sends three packets.

 Click **Menu**, **Accessories**, **Terminal** or click the button on the taskbar to launch a terminal window providing a command prompt

 Enter the command **sudo ping google.com -c 3**, then hit Return to test your internet connection

```
pi@raspberrypi: ~
File  Edit  Tabs  Help
pi@raspberrypi:~ $ sudo ping google.com -c 3
PING google.com (212.39.82.180) 56(84) bytes of data.
64 bytes from 180-82.btc-net.bg (212.39.82.180): icmp_seq=1 ttl=56 time=13.0 ms
64 bytes from 180-82.btc-net.bg (212.39.82.180): icmp_seq=2 ttl=56 time=18.7 ms
64 bytes from 180-82.btc-net.bg (212.39.82.180): icmp_seq=3 ttl=56 time=18.3 ms

--- google.com ping statistics ---
3 packets transmitted, 3 received, 0% packet loss, time 2003ms
rtt min/avg/max/mdev = 13.077/16.707/18.706/2.573 ms
pi@raspberrypi:~ $
```

The results of the **ping** command test show that three 64 byte packets have been sent from your system, and the time taken for each packet to be received by the specified host. A statistical summary confirms that three packets were sent and received without loss, and the total time taken for the test. Additionally, minimum time, mean (average) time, and maximum time are provided for the packets' round-trip, together with a mean deviation figure – a low deviation indicating that the various packet round-trip times are close to the calculated average.

...cont'd

When your Raspberry Pi is connected to the internet you can access the web using the Epiphany web browser, which is bundled with the Raspbian distro. Epiphany is a fast, lightweight browser that uses the WebKit rendering engine, which also powers the Google Chrome and Apple Safari browsers.

The initial home page of the Epiphany browser is a "Most Visited" page, containing thumbnails of recently visited web pages, which you can click to quickly re-open those pages:

If you need assistance with internet connection issues, or indeed any other Raspberry Pi issues, you can find forum help online at **https://www. raspberrypi.org/forums/**

3 Click **Menu**, **Internet**, **Epiphany Web Browser** or click the button on the taskbar to launch the web browser

4 Enter a website URL into the address field, such as **ineasysteps.com**, and hit Return to visit that site

Raspberry Pi has limited ability for web browsing so you may find performance slow on pages containing complex JavaScript.

OMXPlayer is a media player that will utilize Raspberry Pi's GPU acceleration.

Extending the system

The default Raspbian system includes a command-line application named "OMXPlayer" that can be used to play multimedia files, such as MP3 music files and MP4 video files.

When you are using a TV connected to the Raspberry Pi by an HDMI cable, the commands to the **omxplayer** must include a switch of **-o hdmi** to ensure playback through the HDMI cable, e.g. issue the command **omxplayer -o hdmi movie.mp4** at a terminal prompt to play a video file named "movie.mp4".

More usefully, in the GUI, the Raspbian system can be extended to add a menu shortcut, which will execute the omxplayer application whenever you click a media file icon:

 Open a terminal window, then at the prompt enter the command **sudo leafpad** and hit Return to launch the Leafpad text editor with superuser status

```
pi@raspberrypi: ~
File  Edit  Tabs  Help
pi@raspberrypi:~ $ sudo leafpad
```

 Next, precisely type shortcut specifications into Leafpad so it looks exactly like the text in the screenshot below

```
*(Untitled)
File  Edit  Search  Options  Help
[Desktop Entry]
Type=Application
Name=OMXPlayer
Categories=AudioVideo;Player;
Exec=lxterminal --command "omxplayer -o hdmi %f"
Terminal=false
Icon=/usr/share/icons/nuoveXT2/96x96/categories/applications-multimedia.png
```

 Ensure that the details you have typed exactly match those shown above, then click **File**, **Save As** on Leafpad's menu bar – to launch the Save As dialog box

...cont'd

4 Choose "Filesystem" in the Places panel, then navigate to the **/usr/share/applications/** directory and save the text file exactly as **omxplayer.desktop**

You cannot save the file in the **/usr/share/ applications/** directory if you have launched Leafpad without the **sudo** command.

5 Click on the **Menu** button to see that the categories have now been extended with a Sound & Video category including the OMXPlayer. Clicking the OMXPlayer menu item merely launches a brief terminal window, as OMXPlayer is a command-line application, but media files can now be associated with OMXPlayer, to be played by clicking on them or opening from the right-click context menu

OMXPlayer is a command-line application so a terminal window opens when the media is playing – closing that terminal window stops playback, or pressing the **Esc** key.

6 Right-click on a media file and choose **Open with, Sound & Video, OMXPlayer** to associate that file type and to see the file play

Summary

- The default Raspbian desktop provides icon shortcuts for the most popular bundled applications.

- The System Tray contains the Wi-Fi settings button, Volume button, CPU usage monitor, Clock, and Ejecter button.

- The taskbar contains the Menu button, File Manager launcher, Web Browser launcher, and Terminal launcher.

- Preferences can be changed to Customize Look and Feel of window widgets, icons, cursors, and fonts.

- Desktop preferences can be changed to customize the desktop wallpaper, background color, text label font, color and shadow.

- Linux filesystems are arranged in a hierarchical tree below / .

- Most directories beneath the / directory contain system files but **/home** contains user's home directories to store user files.

- File Manager opens by default in the current user's home directory and displays the directory name on its title bar.

- A terminal window provides a prompt that allows commands to be executed as if at the console CLI prompt.

- The **cd** command navigates to a named directory but .. and ~ can be used to nominate parent and username directory.

- Leafpad is a plain text editor where text files can be created.

- All files created by users should be saved in their home directory, or in a sub-directory within their home directory.

- Internet connection can be tested using the **ping** command.

- The Epiphany web browser uses the WebKit rendering engine.

- Multimedia files can be played by the OMXPlayer command-line application.

- The Raspbian system can be extended to allow OMXPlayer to execute playback when you click on a media file icon.

3

Commanding the system

This chapter demonstrates how to issue Linux commands at a terminal or console command prompt.

Listing contents

Issuing commands at a prompt is a powerful way to communicate directly with the operating system, so it's worth becoming familiar with some of the most-used commands, such as the **ls** and **pwd** commands described earlier. Multiple commands can be combined at the prompt if separated by a semi-colon, such as the combined command **pwd ; ls** – although these will be executed as individual consecutive commands.

Another useful command is the **cat** command that will output the text contained in one or more text files in the terminal window.

Commands to work with files include the **cp** command that copies a file to a new specified name, the **mv** command that moves a file to a new specified location (or simply renames a file), and the **rm** command that removes a specified file.

Commands to work with directories include the **cp** command that copies a directory to a new specified name, the **mkdir** command that makes a new directory with a specified name, and the **rmdir** command that removes an empty directory of a specified name.

Use the **man** command to learn more about each command. For example, enter **man cat** to see the manual page for the **cat** command.

The **clear** command has been issued between these screenshots – to clear the previous results from view.

 Launch a terminal window, then issue combined commands to reveal the current directory and list its contents

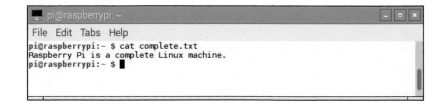

```
pi@raspberrypi: ~
File  Edit  Tabs  Help
pi@raspberrypi:~ $ pwd ; ls
/home/pi
complete.txt  Documents   Music      Public      Videos
Desktop       Downloads   Pictures   Templates
pi@raspberrypi:~ $ ▮
```

 Next, issue a command to display the entire content of the text file within the current directory

```
pi@raspberrypi: ~
File  Edit  Tabs  Help
pi@raspberrypi:~ $ cat complete.txt
Raspberry Pi is a complete Linux machine.
pi@raspberrypi:~ $ ▮
```

3 Now, issue combined commands to make a copy of the text file, list the contents of the current directory to confirm the new file exists, and display the entire content of the new text file

```
pi@raspberrypi: ~
File Edit Tabs Help
pi@raspberrypi:~ $ cp complete.txt raspi.txt ; ls ; cat raspi.txt
complete.txt  Documents  Music      Public      Templates
Desktop       Downloads  Pictures   raspi.txt   Videos
Raspberry Pi is a complete Linux machine.
pi@raspberrypi:~ $ 
```

You can remove all files within the current directory using the * wildcard with the command **rm ***.

4 Issue combined commands to make a new directory, move the new text file into the new directory, list the contents of both the current and new directories – to confirm the new directory exists and now contains the new text file

```
pi@raspberrypi: ~
File Edit Tabs Help
pi@raspberrypi:~ $ mkdir Text ; mv raspi.txt Text ; ls ; ls Text
complete.txt  Documents  Music      Public      Text
Desktop       Downloads  Pictures   Templates   Videos
raspi.txt
pi@raspberrypi:~ $ 
```

5 Issue a command that attempts to remove the new unempty directory – to see the attempt fail

6 Finally, issue combined commands to remove the file within the new directory and that directory itself, then list the contents of the current directory – to see the new directory has now been removed

If you are feeling brave, you can remove a directory and all of its files with the command **rm -r** but use this with care!

```
pi@raspberrypi: ~
File Edit Tabs Help
pi@raspberrypi:~ $ rmdir Text
rmdir: failed to remove 'Text': Directory not empty
pi@raspberrypi:~ $ rm Text/raspi.txt ; rmdir Text ; ls
complete.txt  Documents  Music      Public      Videos
Desktop       Downloads  Pictures   Templates
pi@raspberrypi:~ $ 
```

Getting applications

Applications on all Linux systems are contained in "packages" that are controlled on the system by a package manager. On the Raspbian system this is called the Advanced Package Tool (APT). The package manager maintains a copy on your system of the index of an online repository, from which thousands of applications are available for installation on your Raspberry Pi. APT is a command-line tool that must be run with **sudo** – as superuser status is required to make changes to your system. To ensure that your local copy of the online repository index is up-to-date, simply issue the command **sudo apt-get update**.

You can search for an application of a particular category using the **apt-cache search** command followed by the category you want. This typically returns a list of application names with brief details. When you have selected an application to install, simply issue the command **sudo apt-get install** followed by the application name – the package files get retrieved and the application gets installed:

Use your intuition to guess a likely app name to search the cache for. Here, it seems likely that a screensaver for the X server might be named as "xscreensaver".

1 Launch a terminal window then issue a command to update the local copy of the repository index

```
pi@raspberrypi: ~
File  Edit  Tabs  Help
pi@raspberrypi:~ $ sudo apt-get update
Fetched 9,245 kB in 2min 22s (64.9 kB/s)
Reading package lists... Done
pi@raspberrypi:~ $ 
```

2 Next, issue a command to search for a screensaver for the X window server – guessing a name of "xscreensaver"

There is a delay after each command here while the package list gets refreshed, the cache gets searched, and the application gets installed.

```
pi@raspberrypi: ~
File  Edit  Tabs  Help
pi@raspberrypi:~ $ apt-cache search xscreensaver
xscreensaver - Screensaver daemon and frontend for X11
xscreensaver-data - Screen saver modules for screensaver frontends
xscreensaver-data-extra - Extra screen saver modules for screensaver frontends
xscreensaver-gl - GL(Mesa) screen saver modules for screensaver frontends
xscreensaver-gl-extra - Extra GL(Mesa) screen saver modules
for screensaver frontends
xscreensaver-screensaver-bsod - BSOD screen saver module from XScreenSaver
xscreensaver-screensaver-dizzy - Graphics demo that makes you dizzy
(XScreenSaver hack)
xscreensaver-screensaver-webcollage - Webcollage screen saver module
xss-lock - invoke external screen lock in response to XScreenSaver events
pi@raspberrypi:~ $ 
```

3 Now, issue a command to install the xscreensaver application that you have located

```
pi@raspberrypi: ~
File  Edit  Tabs  Help
pi@raspberrypi:~ $ sudo apt-get install xscreensaver
Reading package lists... Done
Building dependency tree
Reading state information... Done
Setting up xscreensaver-data (5.30-1+deb8u1) ...
Setting up xscreensaver (5.30-1+deb8u1) ...
Processing triggers for libc-bin (2.19-18+deb8u3) ...
Processing triggers for dictionaries-common (1.23.17) ...
pi@raspberrypi:~ $
```

If you get asked if you wish to **continue [Y/n]** during the installation, just press the Y keyboard key then hit Return.

4 Click the newly added **Menu**, **Preferences**, **Screensaver** menu item to launch the Screensaver Preferences dialog

Menu

- Programming ›
- Office ›
- Internet ›
- Games ›
- Accessories ›
- Sound & Video ›
- Help ›
- Preferences › — Add / Remove Software
- Run... — Appearance Settings
- Shutdown... — Audio Device Settings
- Main Menu Editor
- Mouse and Keyboard Settings
- Raspberry Pi Configuration
- Screensaver

5 Choose your screensaver options to see a preview, then close the Screensaver Preferences dialog to apply your choices

Screensaver Preferences (XScreenSaver 5.30)

File Help

Display Modes | Advanced

Mode: Random Screen Saver

FuzzyFlakes

- ☐ Flurry
- ☐ FlyingToasters
- ☑ FontGlide
- ☑ FuzzyFlakes
- ☑ Galaxy
- ☐ Gears
- ☐ Geodesic
- ☐ GFlux

Blank After 10 minutes
Cycle After 10 minutes
☐ Lock Screen After 0 minutes

Preview   Settings...

Use **sudo apt-get remove** followed by an application name to uninstall that app.

43

The **wget** command has a useful **-c** switch option that restarts a stopped download from the point where it was halted.

The **-E** switch ensures that server-generated **text/html** files with extensions like **.asp** will be converted for local viewing.

Saving web pages

The Raspbian system has a powerful **wget** command-line application for saving web pages and files onto your computer. This command in its simplest form merely requires you to specify the URL of a page or file to be saved to the current directory. For example, the command **wget www.example.com/code.zip** will download a copy of the specified ZIP archive to the current directory, while displaying a progress and this information:

- **% of download complete** – e.g. 35%
- **Total of bytes downloaded so far** – e.g. 22,400
- **Current download speed** – e.g. 88.5K/s
- **Remaining time to completion** – e.g. eta 12s

The command **wget www.example.com** will typically store that domain's base **index.html** file (home page) on your local system – but not any support files it may use, such as image files or stylesheet files. This might be what you need, but often you will also want to retrieve the support files to save a complete local copy of the web page. Fortunately, the **wget** command has lots of switch options to help you save precisely what you want:

- **-E** – adjust Extensions for files of type **text/html** to **.html**
- **-H** – enable Host spanning to retrieve support files
- **-k** – convert links in the document for local viewing
- **-K** – bacKup original version of converted files with **.orig** suffix
- **-p** – acquire page requisites such as image files
- **-P** – directory Prefix where all retrieved files should be placed

A command using all these switches might look like this:
wget -E -H -k -K -p www.example.com -P SavedPageDirectory

The web page shown at the top of the page opposite comprises an HTML document supported by an adjacent JavaScript file for functionality and an adjacent CSS stylesheet for presentation. Additionally, it has a HTML5 logo image from the W3C server.

...cont'd

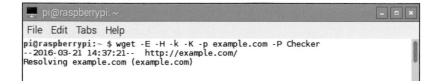

The icon in the browser address field here is not part of the document, so will not be saved in the local version of the page.

 Enter a command to save the web page and all its support files in a new directory named "Checker"

```
pi@raspberrypi: ~
File Edit Tabs Help
pi@raspberrypi:~ $ wget -E -H -k -K -p example.com -P Checker
--2016-03-21 14:37:21--  http://example.com/
Resolving example.com (example.com)
```

 When saving is complete, disconnect from the internet

3 Now, open the saved web page in the Epiphany browser to see that CSS presentation, JavaScript functionality, and remote host images are all preserved locally

You can try to download an entire website using a command something like **wget -r -k -U Mozilla example.com**

Reading & writing text

The Raspbian system provides a useful command-line editor named "nano". This runs in the terminal window from which it gets launched, and returns to a command prompt when closed. Plain text files can therefore be quickly created or existing ones edited directly using only the terminal window.

 Enter the command **nano** to launch the text editor in the terminal window, then type in some simple text

 Press the **Ctrl + X** keys on your keyboard to exit nano

When typing text in nano press the Return key to insert a line return in the text.

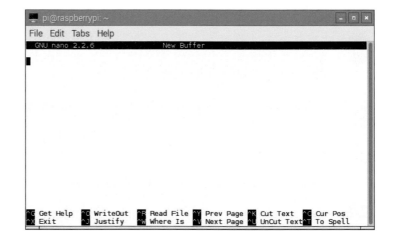

 Press the **Y** key on your keyboard to save the text you have typed into nano

The nano options use a ^ carat character to mean the Ctrl key – for example, press Ctrl + G keys for help.

4 Type the name of a file to save this text in as **simple.txt** then hit Return to write the file and exit nano

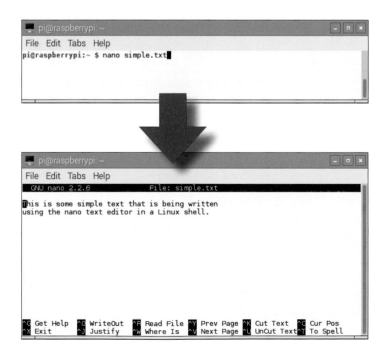

Notice that the available options at the bottom of the nano window change at each stage.

5 To reopen an existing text file in the text editor at any time, simply type **nano** followed by the file name or location address, then hit Return

An alternative command-line editor to nano is the vi text editor – enter **vi** at a prompt and hit Return to launch it in the terminal window.

Finding files

Locating a file on your Raspbian system can be achieved from a shell prompt using the **find** command. This is a very powerful command, with over 50 possible options, but it has an unusual syntax. Possibly the one most used looks like this:

find *DirectoryName* **-type f -name** *FileName*

The directory name specifies the hierarchical starting point from which to begin searching. If you know the file exists somewhere in your home directory structure, you could begin searching there (**~**). Alternatively, you could use the **sudo** command to assume root superuser status then search the entire file system by starting from the top-level root directory (**/**).

In this case, the **-type f** option specifies that the search is for a file – denoted by the letter "f". The **-name** option makes the search by name – seeking the specified file name.

Use the wildcard * with the file name when you know the name but not the extension.

 Launch a terminal window, then at the prompt enter the command **find ~ -type f -name simple.txt** to seek all files named **simple.txt** in your home directory structure

```
pi@raspberrypi: ~
File  Edit  Tabs  Help
pi@raspberrypi:~ $ find ~ -type f -name simple.txt
/home/pi/simple.txt
pi@raspberrypi:~ $
```

 Assume superuser status using **sudo** followed by **find / -type f -name simple*** to search the entire file system for any file whose name begins with "simple"

```
pi@raspberrypi: ~
File  Edit  Tabs  Help
pi@raspberrypi:~ $ sudo find / -type f -name simple*
/home/pi/simple.txt
pi@raspberrypi:~ $
```

You can recall the last command entered by pressing the up arrow key.

A so-called "hard link" to a file can be created with the **ln** command – producing a link pointing to the system address of the file. Typically, this might be used to place a shortcut on your desktop to a file somewhere within your home directories:

3 Enter a command to display the current directory and contents then a command to create a desktop hard link

linkto-complete.txt

```
pi@raspberrypi: ~
File Edit Tabs Help
pi@raspberrypi:~ $ pwd ; ls
/home/pi
complete.txt  Documents  Music     Public     Templates
Desktop       Downloads  Pictures  simple.txt  Videos
pi@raspberrypi:~ $ ln complete.txt Desktop/linkto-complete.txt
pi@raspberrypi:~ $
```

By default, the **find** command will only report the location of the actual file but you can also have it include hard links in the report by adding a **-samefile** option:

 4 Enter a command to find a file and any hard links to it

```
pi@raspberrypi: ~
File Edit Tabs Help
pi@raspberrypi:~ $ find ~ -samefile complete.txt
/home/pi/complete.txt
/home/pi/Desktop/linkto-complete.txt
pi@raspberrypi:~ $
```

A hard link is essentially a copy of the file that is not automatically removed if the original file gets deleted – so it is useful to find hard links then delete all versions.

49

In addition to searching for files, the **-type** option can specify that the search is for directories – denoted by the letter "d". When searching directory structures of many levels, it may be desirable to limit the search depth with the **find** command's **-maxdepth** option to specify the number of levels to search.

5 Enter **find /usr -maxdepth 1 -type d** to report only immediate sub-directories of the **/usr** directory

```
pi@raspberrypi: ~
File Edit Tabs Help
pi@raspberrypi:~ $ find /usr -maxdepth 1 -type d
/usr
/usr/share
/usr/bin
/usr/src
/usr/include
/usr/local
/usr/games
/usr/lib
/usr/sbin
pi@raspberrypi:~ $
```

The **/usr** directory has many levels – remove the -maxdepth 1 option from the command in Step 5 and run the command again to see all directories.

Adding users

User accounts are administered by the superuser who can issue a **useradd** command followed by a username to create a new user account. This command has a **-m** switch to create a username directory for that new user when the account gets created.

After creating a new user account, the superuser can create an initial password for that account with the **passwd** command. Finally, the superuser can provide a login prompt with the **login** command so a user can log in by entering their username and password. As usual, temporary superuser status can be assumed by prefixing each of these actions with the **sudo** command.

 Enter the **sudo** command followed by **useradd -m mike** to create a new user "mike" and a username directory

 Next, enter the command **sudo passwd mike** to get a new password prompt for the new user "mike" – then type that user's initial password twice

 Now, enter the command **sudo login** to get a new login prompt at which any user can log in

 Enter the username "mike" and password to log in then enter the command **pwd** to see that user is automatically logged in within their new home directory

 Enter a **logout** command to return to the previous user's prompt and filesystem location

If you omit the **-m** switch from **useradd**, the user will be created without any place to store files they create.

Use the commands **sudo userdel** and **sudo groupdel** followed by a name to delete users and groups respectively.

```
pi@raspberrypi: ~
File  Edit  Tabs  Help
pi@raspberrypi:~ $ sudo useradd -m mike
pi@raspberrypi:~ $ sudo passwd mike
Enter new UNIX password:
Retype new UNIX password:
passwd: password updated successfully
pi@raspberrypi:~ $ sudo login mike
Password:
mike@raspberrypi:~ $ pwd
/home/mike
mike@raspberrypi:~ $ logout
pi@raspberrypi:~ $ pwd
/home/pi
pi@raspberrypi:~ $
```

A group is a set of user accounts whose rights can be modified simultaneously. The root superuser might, for instance, grant a group permission to access a previously inaccessible file – all users who are members of that group are then allowed access.

Any user can discover which groups they belong to with the **groups** command. The superuser can append a username after the command to reveal the group membership of that particular user.

The superuser can also add a new group with a **groupadd** command, followed by a new group name, and add a user to an existing group with a command **usermod -G** followed by the group name and username.

Once again, temporary superuser status is assumed by prefixing each of these actions with the **sudo** command.

You must make a user a member of the **sudo** group to allow them to use the **sudo** command.

 At a user prompt, enter the command **groups** to discover the groups of which this user is a member

 Now, enter the command **sudo groups mike** to discover the groups of which the user "mike" is a member

 Issue the command **sudo groupadd band** to create a new group named "band"

 Now, enter a command to make "mike" a member of two groups **sudo usermod -G band,sudo mike**

 Enter **sudo groups mike** to see all groups of which this user is now a member

After creating a new user, you can click the Logout button in the System Tray then log back in with the new username and password.

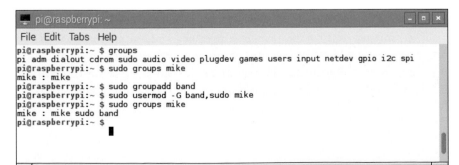

```
pi@raspberrypi: ~
File  Edit  Tabs  Help
pi@raspberrypi:~ $ groups
pi adm dialout cdrom sudo audio video plugdev games users input netdev gpio i2c spi
pi@raspberrypi:~ $ sudo groups mike
mike : mike
pi@raspberrypi:~ $ sudo groupadd band
pi@raspberrypi:~ $ sudo usermod -G band,sudo mike
pi@raspberrypi:~ $ sudo groups mike
mike : mike sudo band
pi@raspberrypi:~ $
```

You can inspect and modify your own files' permissions with File Manager – right-click on a file icon then choose Properties, Permissions.

Notice that the "mike" user is a member of the group named "band", which was created on the previous page, and the "sudo" group – so the **sudo** command is enabled for this user.

Changing permissions

The **ls -l** long listing command reveals the access permissions of each item in the current directory as a string of 10 characters at the beginning of each line. The first is a **d** for a directory, or a dash for a file. This is followed by sequential Read, Write and Execute permissions for the owning User, Group, and Others. Characters **r**, **w**, and **x** appear for those permissions that are set, otherwise a dash is shown.

In the listing below, a script may be Read and Executed by everyone and the owner (this user) may also Write to it. The owner can both Read and Write a text note but others may only Read it. Three other text files may only be Read or Written to by their respective owners – so this user cannot access them.

```
mike@raspberrypi: ~
File  Edit  Tabs  Help
mike@raspberrypi:~ $ whoami
mike
mike@raspberrypi:~ $ ls -l
total 20
-rw-------  1 andy andy   36 Mar 21 16:06 andys-note
-rw-------  1 dave dave   44 Mar 21 16:06 daves-note
-rw-------  1 mike band   39 Mar 21 15:51 mikes-note
-rw-r--r--  1 tony tony   28 Mar 21 16:07 tonys-note
-rwxr-xr-x  1 tony tony  326 Mar 21 16:07 tonys-script
mike@raspberrypi:~ $ cat tonys-note
Hi! From your buddy Tony...
mike@raspberrypi:~ $ cat daves-note andys-note mikes-note
cat: daves-note: Permission denied
cat: andys-note: Permission denied
Here are Mike's thoughts of the day...
mike@raspberrypi:~ $ 
```

Each set of permissions can also be described numerically where Read = 4, Write = 2 and Execute = 1. For instance, a value of 7 describes full permissions to Read, Write and Execute (4 + 2 + 1), 5 describes permissions to Read, Execute (4 + 1), and so on.

Permissions can be changed at a shell prompt with the **chmod** command, stating the permission values and the file name as its arguments. For example, the command **chmod 777 myfile** sets full permissions for a file named "myfile" in the current directory. If you need to change permissions where you are not the owner, you must assume root superuser status with the **sudo** command.

The **chgrp** command can be used to change the group membership of a file by stating a group name and the file name as its arguments. Similarly, the **chown** command can specify a username and the file name to change the user ownership. Temporary superuser status is assumed by prefixing each of these with **sudo**.

Do not fall into the habit of setting everything to permissions of 777 – use access permissions thoughtfully to maintain useful restrictions.

 1 Enter the command **sudo chmod 644 mikes-note** to additionally allow everyone to read this file

 2 Next, enter the command **chmod 640 daves-note** to allow the group to read this file

3 Enter the command **sudo chgrp band daves-note** to change the group to one of which this user is a member – so this user may now read this file

 4 Now, enter the command **sudo chown mike andys-note** to change ownership of this file – so this user may now read the file

5 Issue an **ls -l** command to see the changes, then read the files where access was previously denied

```
mike@raspberrypi: ~
File  Edit  Tabs  Help
mike@raspberrypi:~ $ sudo chmod 644 mikes-note
mike@raspberrypi:~ $ sudo chmod 640 daves-note
mike@raspberrypi:~ $ sudo chgrp band daves-note
mike@raspberrypi:~ $ sudo chown mike andys-note
mike@raspberrypi:~ $ ls -l
total 20
-rw-------  1 mike andy   36 Mar 21 16:06 andys-note
-rw-r-----  1 dave band   44 Mar 21 16:06 daves-note
-rw-r--r--  1 mike band   39 Mar 21 15:51 mikes-note
-rw-r--r--  1 tony tony   28 Mar 21 16:07 tonys-note
-rwxr-xr-x  1 tony tony  326 Mar 21 16:07 tonys-script
mike@raspberrypi:~ $ cat daves-note andys-note
Here are some pearls of wisdom from Dave...
Hey, here's a greeting from Andy...
mike@raspberrypi:~ $
```

A file can easily be made executable using the command **chmod +x** – short for **chmod 711**.

The arithmetic operators are **+** add, **-** subtract, ***** multiply, **/** divide, and **%** modulus.

The command-line language in your Raspbian distro is called Bash – an acronym of Bourne Again SHell.

Employing the shell

The command-line in Linux is known as the "shell" and can be used to perform simple math calculations at a command prompt with the **expr** command. This recognizes all the usual arithmetic operators, but those which have other meanings in the shell need to be prefixed by a backslash \ escape character. For instance, the * wildcard must be escaped for multiplication. Each argument must be separated by whitespace, and parentheses can be used to establish operator precedence in longer expressions – but each parenthesis character must be escaped.

The **expr** command can also perform Boolean evaluations that return either true (1) or false (0) answers, and perform simple string manipulation with functions **length**, **substr**, and **index**.

 At a shell prompt, enter **expr 7 + 3** to perform addition and **expr 7 * 3** to perform multiplication

 Enter **expr 7 * \\(3 + 1 \\)** to evaluate a complex expression and **expr 7 = 3** to evaluate equality

 Issue the command **expr length "Raspberry Pi in easy steps"** to discover the length of the specified string

 Issue the command **expr substr "Raspberry Pi in easy steps" 14 13** to extract a substring of the specified string

 Issue the command **expr index "Raspberry Pi in easy steps" "b"** to find the string position of the first "b"

```
pi@raspberrypi: ~                                          _ □ x
File  Edit  Tabs  Help
pi@raspberrypi:~ $ expr 7 + 3
10
pi@raspberrypi:~ $ expr 7 \* 3
21
pi@raspberrypi:~ $ expr 7 \* \( 3 + 1 \)
28
pi@raspberrypi:~ $ expr 7 = 3
0
pi@raspberrypi:~ $ expr length "Raspberry Pi in easy steps"
26
pi@raspberrypi:~ $ expr substr "Raspberry Pi in easy steps" 14 13
in easy steps
pi@raspberrypi:~ $ expr index "Raspberry Pi in easy steps" "b"
5
pi@raspberrypi:~ $ ▮
```

...cont'd

The result of an expression evaluation can be made to cause a particular action using an **if-then-else** statement. This has three separate parts that specify a test expression, the action to perform when the test is true, and the action to perform when it is false.

The **if** keyword begins the statement and is followed by the test expression enclosed within a pair of **[ ]** square brackets. Each part of the entire statement must be separated from the next by whitespace to enable the shell to evaluate the expression.

The **then** keyword begins the second part of the statement, specifying the commands to execute when the test is true. Similarly, the **else** keyword begins the third part of the statement, specifying the commands to execute when the test is false. Finally, the **fi** keyword must be added to mark the end of the statement.

You may type the first part of the statement then hit Return to be prompted to enter the rest of the statement, or type the entire statement separating each part with a semicolon:

 Type **if [`expr 7 % 2` = 0]** to test if the remainder of dividing seven by two is zero, then hit Return

 At the prompt, enter **then echo "Even number"**

 Now, enter **else echo "Odd number"**

9 Type **fi** then hit return to perform the appropriate action

10 Make a similar test in a continuous statement by typing **if [`expr 8 % 2` = 0]; then echo "Even Number"; else echo "Odd Number"; fi Even Number** then hit Return to perform the appropriate action

The backtick ` operators enclose **expr 7 % 2** so that operation gets performed before the test expression is evaluated – in this case the remainder of one is substituted, making the expression **if [1 = 0]**.

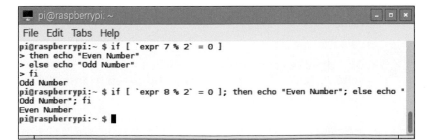

```
pi@raspberrypi: ~
File  Edit  Tabs  Help
pi@raspberrypi:~ $ if [ `expr 7 % 2` = 0 ]
> then echo "Even Number"
> else echo "Odd Number"
> fi
Odd Number
pi@raspberrypi:~ $ if [ `expr 8 % 2` = 0 ]; then echo "Even Number"; else echo "
Odd Number"; fi
Even Number
pi@raspberrypi:~ $
```

Creating shell scripts

Lengthy shell routines, like those on the previous page, can be conveniently saved as a shell script for execution when required. Shell scripts are simply plain text files that begin their first line with **#!/bin/bash**, specifying the location of the bash program, and are typically given a **.sh** file extension.

Once a shell script file has its access permission set to "executable", it can be executed at any time by prefixing its name or path with the **./** dot-slash characters at a shell prompt.

A script might, perhaps, employ the **$RANDOM** shell variable that generates an integer from zero to 32,767 each time it is called. These are not truly random, however, as the same sequence is generated given the same starting point (seed) – in order to ensure different sequences, it is necessary to set it with different seeds.

One solution is to extract a dynamic value from the current time using the **date +%s** command to deliver the current number of seconds that have elapsed since **00:00:00 GMT January 1, 1970**. Using this to seed the **$RANDOM** variable gets better random number generation.

Arithmetic can be performed on shell variables, such as **$RANDOM**, by including the **let** command in a script.

Enter **RANDOM=1** then **echo $RANDOM** three times to see the pattern. Repeat the commands to see the pattern repeat.

56

You must not introduce whitespace around the = character in the assignments.

 Open any plain text editor and begin a new file with the line **#!/bin/bash**

 On a new line, type a line to seed the **$RANDOM** variable
RANDOM=`date +%s`

 Add a line to assign a value 1-20 to a variable
let NUM=($RANDOM % 20 + 1)

 On the next line, type an instruction to clear the window
clear

 Add these lines to output text for the user
echo "I have chosen a number between 1 and 20"
echo "Can you guess what it is?"

6 Now, add a line to read the user's guess into a variable
read GUESS

7 Type the following lines exactly to evaluate whether the user's guess matches the generated number and output an appropriate response for each incorrect attempt
while [$GUESS -ne $NUM]
do

      if [$GUESS -gt $NUM]
          then echo "No - try lower... "
          else echo "No - try higher... "
      fi
      read GUESS
done

The **while-do** statement is a loop that employs the **-ne** (not equal) comparison operator and a **-gt** (greater than) comparison operator. Other Bash shell comparison operators include **-eq** (equal) and **-lt** (less than).

8 Add a line to confirm a correct guess
echo "Yes the number is $NUM"

9 Save the script as **guess.sh** in the current directory

10 At a shell prompt, change the access permissions to make the script executable by its owner with this command
chmod 711 guess.sh

11 Enter **./guess.sh** to execute the script

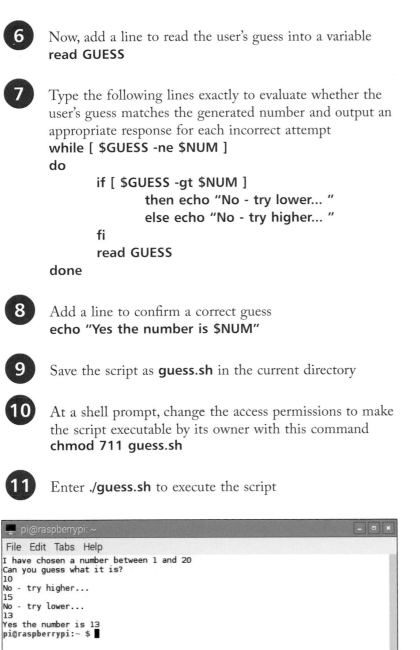

```
pi@raspberrypi: ~
File Edit Tabs Help
I have chosen a number between 1 and 20
Can you guess what it is?
10
No - try higher...
15
No - try lower...
13
Yes the number is 13
pi@raspberrypi:~ $
```

Remember that the square bracket characters are operators – there must be space around them to avoid errors.

Summary

- Multiple commands can be combined at a prompt if separated by a semi-colon.
- The **cat** command outputs text from one or more text files in the terminal window.
- The **cp** command copies a file, the **mv** command moves or renames a file, and the **rm** command removes a file.
- The **mkdir** command makes a new directory and the **rmdir** command removes an empty directory.
- APT is a command-line tool that must be run with **sudo**.
- The local repository index gets updated with **apt-get update** and can be searched with the **apt-cache search** command.
- Applications can be added with the **apt-get install** command and removed using the **apt-get remove** command.
- Complete web pages can be saved with **wget -E -H -k -K -p**.
- The **nano** plain text editor runs inside a terminal window.
- Any file on your system can be located with the **find** command.
- User accounts are administered by the superuser with **sudo**.
- The **useradd -m** command creates a user and username directory then the **passwd** command can supply a password.
- The **sudo login** command provides a user login prompt.
- The superuser can add a new group with **groupadd** and add a user to an existing group with the **usermod -G** command.
- The **chmod** command changes access permissions, **chgrp** changes group membership, and **chown** changes ownership.
- The **expr** command allows math calculations and Boolean evaluations to be performed at the terminal shell prompt.
- Shell scripts are text files that begin with **#!/bin/bash** to state the location of the bash program and have a **.sh** file extension.

4 Animating with Scratch

This chapter demonstrates

how to create animations

with the Scratch application.

Walking a sprite

The Scratch application, which is included with the Raspbian distro, is a fun learning environment that enables complete beginners to create computer programs visually – without writing any code. Scratch is intended to encourage an interest in programming by allowing the novice to easily create interactive animation and games.

Scratch programs are built by connecting a number of jigsaw-puzzle shaped blocks to make a character perform some action. Each block has a label, describing its purpose, and the blocks get connected simply by drag 'n' drop within the Scratch interface.

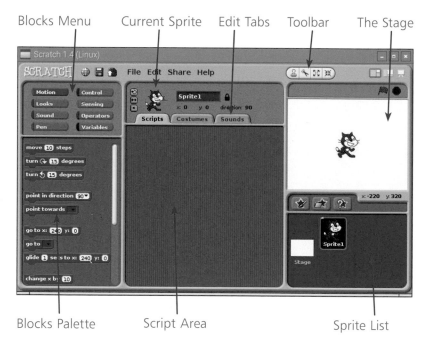

Blocks Menu   Current Sprite   Edit Tabs   Toolbar   The Stage

Blocks Palette   Script Area   Sprite List

The Blocks Menu selection changes the Blocks Palette, making a different selection of blocks available. Dragging any block from the palette onto the Script Area leaves a copy of that block in the Script Area. Dropping any compatible block onto another snaps them together, connecting them like jigsaw-puzzle pieces.

The Scratch interface can be launched by clicking **Menu**, **Programming**, **Scratch**. Initially, a cat character sprite appears on The Stage and can be animated simply by connecting two blocks:

Blocks from the Control and Motion menus are used initially for animation – but go ahead and explore blocks on the other menus.

1   Select the **Control** menu, then drag the "when space key pressed" block from the Palette onto the Script Area

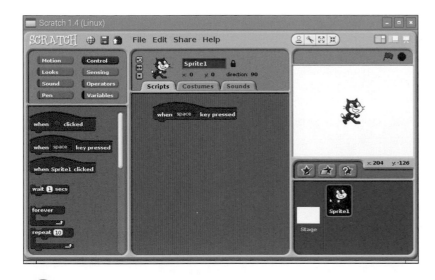

You can drag blocks around the Script Area to connect or disconnect them at any time.

2   Now, select the **Motion** menu, then drag the "move 10 steps" block from the Palette onto the Script Area – connecting it to the first control block

3   Repeatedly press the space key on your keyboard to see the cat sprite character walk across The Stage

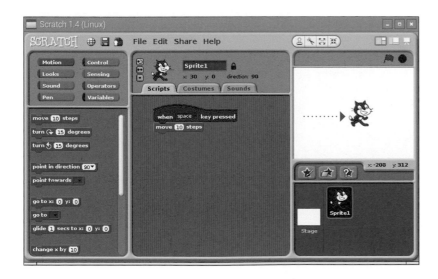

You can drag the cat around The Stage to reposition the sprite.

Changing directions

Building on the example on the previous page, which created a program to walk a sprite forward, you can extend the program to allow the sprite to walk in different directions:

 1 In the Script Area click on the drop-down button on the "when space key pressed" block, then choose "right arrow" from the menu – to change its label so it now reads "when right arrow key pressed"

You can use **File**, **Save** on the Scratch menu to store your animation program at any time.

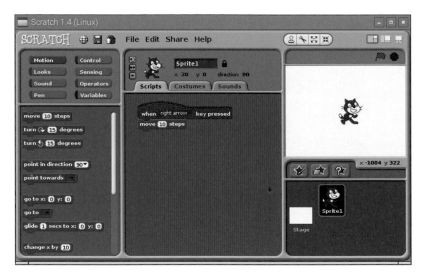

 2 Select the **Control** menu, then drag another "when space key pressed" block from the Palette onto the Script Area

 3 In the Script Area, click on the drop-down button on this "when space key pressed" block, then choose "left arrow" from the menu – to change its label so it now reads "when left arrow key pressed"

You can drag blocks around the Script Area to create more empty space in which to drop further blocks.

 4 Now, select the **Motion** menu, then drag another "move 10 steps" block from the Palette onto the Script Area – connecting it to the second control block

 5 In the Script Area, click on the number on this "move 10 steps" block, then edit its value to -10 – to change its label so it now reads "move -10 steps"

...cont'd

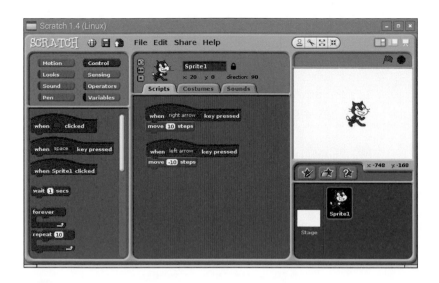

The X axis relates to horizontal movement, and the Y axis relates to vertical movement.

6 Drag in two more "when pressed" control blocks and set them to control with "up arrow" and "down arrow" keys

7 Drag in two "move y by" pieces and set their values to 10 and -10 – connecting them to the last two control blocks

8 Press the arrow keys on your keyboard to see the cat sprite character walk around The Stage, changing direction

63

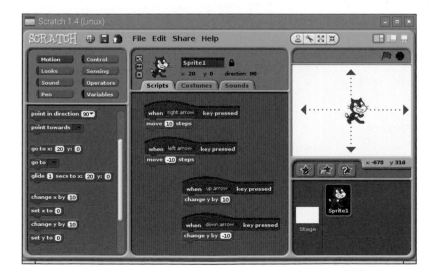

There are often several ways to produce a particular action in Scratch – "move x by 10" is equivalent to "move 10 steps".

Adding another sprite

Building on the example on the previous page, which enabled a sprite to walk in four directions, you can extend the program by adding another sprite to interact with the first one:

 In the Sprite List, click on the ⭐ button to launch the New Sprite dialog, then choose the **Animals** category

The sprites are simply image files in PNG format located in **/usr/share/scratch/Media/Costumes/Animals** – you can add your own sprites to your animations.

 In the New Sprite dialog box, choose a dog sprite then click the **OK** button to see it appear on The Stage

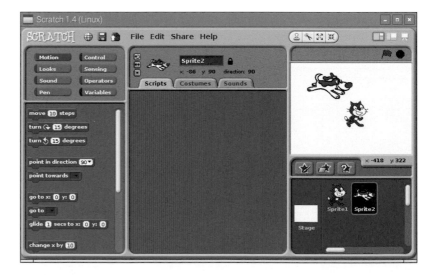

Select any sprite icon in the Sprite List to see the Script Area display the instruction blocks for that selected sprite.

The Script Area is empty, as you have not yet created any instructions to control movement of the dog sprite.

3 Select the **Control** menu, then drag a "when clicked" block from the Palette onto the Script Area

4 Next, drag a "forever" control block onto the Script Area – connecting it to the first control block

5 Now, select the **Motion** menu, then drag a "point towards" block from the Palette onto the Script Area – connecting it to the "forever" control block

6 Click the arrow on the "point towards" block and choose "Sprite 1" from the drop-down menu – so the dog sprite will always point towards the cat sprite

7 Drag a "move 10 steps" motion block onto the Script Area – connecting it below the first motion block

8 Finally, change the value to 1 on the "move 10 steps" motion block – so the dog will move slower than the cat

9 Click the button to play the animation, then use the arrow keys to have the cat outrun the chasing dog

Hot tip

Use the right-hand button on the Scratch Toolbar to reduce the size of the sprites on The Stage – so the cat can easily outrun the dog.

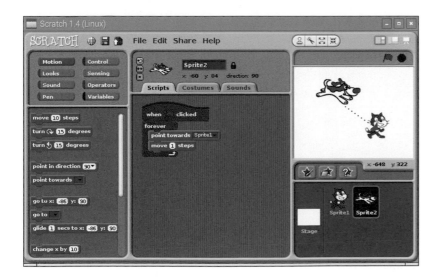

Don't forget

Use the red button to stop the animation, and the mouse to drag the sprites around The Stage.

Editing costumes

Building on the example on the previous page, which added a second sprite, you can develop the program by editing the appearance of the sprites:

 Select "Sprite 1" in the Sprites List, then select the **Costumes** tab and click the **Edit** button beside the current costume – to launch the Paint Editor dialog

Notice that the cat has a second, similar costume that could be used for animation by alternating between the two.

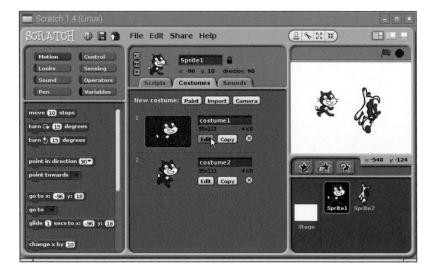

Experiment with the Paint Editor tools as much as you like – you can always import the cat costume to restore the original appearance.

Paint Editor tools can be used to change the appearance of the current costume by editing colors, shape and size, or you can use the **Import** button to choose a new costume.

2 Click the **Import** button, then choose a crab costume from the dialog that appears – the new costume then gets added to the Paint Editor dialog

3 Drag the cursor around the cat costume to select it, then hit the **Delete** button on your keyboard to remove it

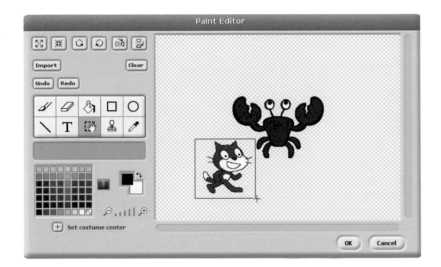

If you don't remove the cat costume, the cat and crab will both appear on The Stage when you close the Paint Editor.

4 Edit the new costume if you wish, then click **OK** to close the Paint Editor and see the new crab costume on The Stage

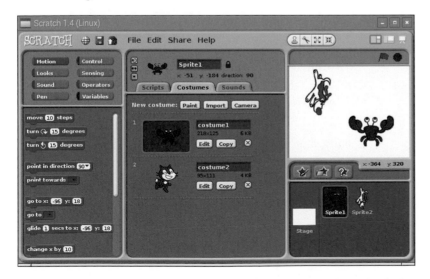

The crab has two similar costumes, as does the cat – you can edit Costume 2 here to import the other crab costume.

Turning around

Building on the example on the previous page, which changed the sprite appearance, you can develop the program by enabling sprite rotation for more realistic movement:

 1 Select "Sprite 1" in the Sprite List, then drag all blocks back onto the Palette except those control blocks for when up arrow, left arrow, right arrow keys are pressed

If you drag blocks off the Script Area and drop them outside the Palette, they will float back onto the Script Area.

68

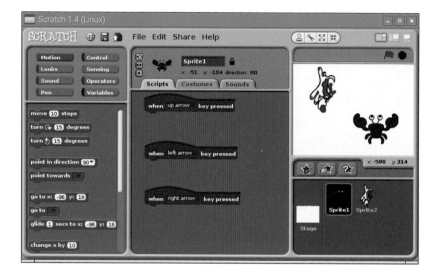

 2 Next, select the **Motion** menu, then drag a "move 10 steps" block from the Palette onto the Script Area – connecting it to the "when up arrow key pressed" control block

3 Now, drag a "turn 15 degrees" counter-clockwise block from the Palette onto the Script Area – connecting it to the "when left arrow key pressed" control block

4 Finally, drag a "turn 15 degrees" clockwise block from the Palette onto the Script Area – connecting it to the "when right arrow key pressed" control block

You can edit the number of degrees to rotate by on each turn block.

...cont'd

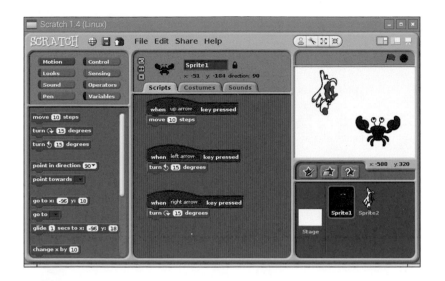

The down button is not required as the sprite will not walk backwards.

5 Click the 🏴 button to play the animation, then press the up arrow key to make the crab walk and use the left and right arrow keys to rotate the crab – try to evade the dog

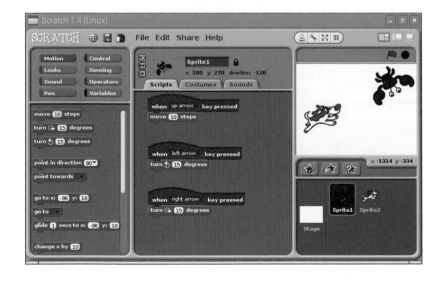

Remember that the up arrow moves the crab 10 steps sideways, not forward – because crabs walk sideways.

Reacting to touch

Building on the example on the previous page, which enabled the sprite to rotate, you can develop the program by making a sprite react when touched by another sprite:

The "when clicked" block simply means "when animation plays".

1 Select "Sprite 1" and the **Control** menu, then drag a "when clicked" block from the Palette onto the Script Area

2 Drag a "forever" block from the Palette onto the Script Area – connecting it to the "when clicked" block

3 Next, drag an "if" block from the Palette onto the Script Area – connecting it to the "forever" block

The "forever" block creates a loop – statements connected within the "forever" block are continually executed when the animation is playing.

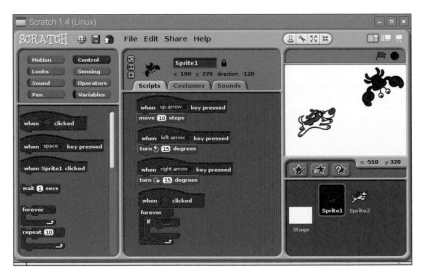

4 Now, select the **Sensing** menu and drag a "touching" block onto the Palette and choose "Sprite 2" from the drop-down menu – connecting it to the "if" block

5 Finally, select the **Looks** menu, then drag a "say Hello! for 2 secs" block onto the Palette and change its value to one second – connecting it to the "touching" block

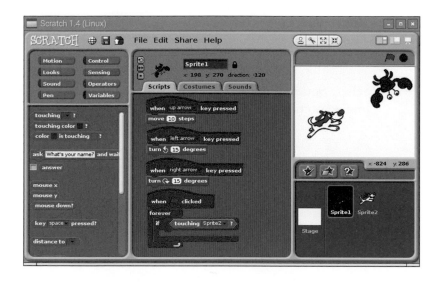

In programming terms, the Sensing blocks test for a "Boolean" value of True or False – "touching" is True when the sprites are touching, otherwise it is False.

6 Click the ![flag] button to play the animation, then press the up arrow key to make the crab walk and use the left and right arrow keys to rotate the crab – but see the message appear when the sprites touch

In programming terms, the "if" block performs a conditional test – so the message will display if the sprites are touching.

Playing sounds

Building on the example on the previous page, which enabled the sprite to react visually when touched by another sprite, you can add sound to play whenever the sprites touch:

 Select "Sprite 1" and the **Sounds** tab, then click the **Import** button to launch the Import Sound dialog

 Choose the "Ya" sound in the **Vocals** category, then click **OK** to see that sound get added to the Sounds Panel

Click on the items in the Import Sound dialog to hear a preview play.

Click on the **Record** button to create your own sound to be played.

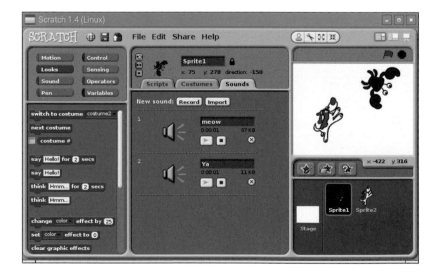

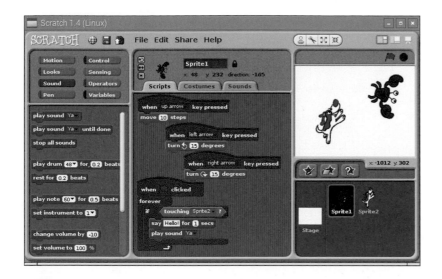

Click on the blocks in the Sound Palette to hear them play a preview.

3 Select the **Sound** menu, then drag a "play sound" block onto the Script Area – connecting it to the "if" block

4 Set both the "say" and "play" blocks to "Ya" on their drop-down menus, then play the animation to hear the sound play when the sprites touch

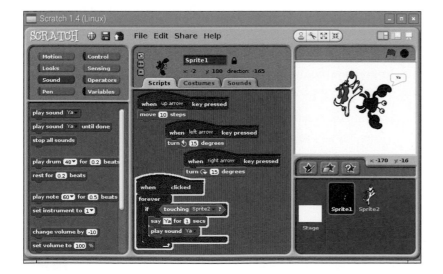

The sound will repeatedly play while the sprites remain touching.

Summary

- Scratch is a learning environment that enables complete beginners to create computer programs visually.

- Each menu in Scratch provides blocks that can be connected to make a sprite character perform an action.

- The "when key pressed" control block detects when a keyboard key is pressed and can be made to move a character.

- The arrow keys can be used to walk a sprite character in various directions when pressed.

- New ready-made sprites can be added to the animation from categories in the New Sprite dialog.

- A sprite can be made to follow another sprite using a "point towards" Motion block.

- The appearance of sprite characters can be changed using the Paint Editor from within the Costumes tab.

- Costumes can be imported to change the appearance of a sprite using the Import button within the Costumes tab.

- Sprite characters can be made to rotate in the animation for more realistic movement using "turn" Motion blocks.

- An animation can react when sprite characters are touching by displaying a message using a "say" Looks block.

- Sounds can be added to the animation from categories in the Import Sound dialog within the Sounds tab.

- The "forever" block creates a programming loop that continually executes the statements on any blocks it contains.

- The "if" block performs a conditional test, so will only execute the statements on any blocks it contains if the test succeeds.

- The "touching" block and other blocks in the Sensing Palette test for a Boolean value of True or False.

5 Programming with Python

This chapter demonstrates how to create computer programs with the Python programming language.

Discover all the latest Python news online at **www.python.org**

The default version of Python on Raspberry Pi is Python 2.7, which gets started with the **python** command. Python 3 is also supplied, which gets started with the **python3** command. All examples in this book use the default of Python 2.7.

Introducing the interpreter

Raspberry Pi is intended to encourage an interest in computer programming and its preferred programming language is Python. This is a high-level (human-readable) programming language that is processed by the Python "interpreter" to produce results fast. Python includes a comprehensive standard library of tested code modules that can be easily incorporated into your own programs.

The Python interpreter compiles text-based program code (script) and also has an interactive mode that is started from a command prompt, simply by entering the command **python**. This will respond with version information, then produce the Python primary prompt **>>>** where you can interact with the interpreter:

 Launch a terminal window, then enter **python** at the command prompt to start Python interactive mode

 At the Python prompt, type a simple addition sum, then hit Return to see the interpreter return the sum total

```
File Edit Tabs Help
pi@raspberrypi:~ $ python
Python 2.7.9 (default, Mar  8 2015, 00:52:26)
[GCC 4.9.2] on linux2
Type "help", "copyright", "credits" or "license" for more information.
>>> 8 + 4
12
>>>
```

Parentheses can be used to group expressions, so their meaning is made clear to the interpreter:

 Enter a sum with three components, then enter it again using parentheses to group the components for clarity

```
File Edit Tabs Help
pi@raspberrypi:~ $ python
Python 2.7.9 (default, Mar  8 2015, 00:52:26)
[GCC 4.9.2] on linux2
Type "help", "copyright", "credits" or "license" for more information.
>>> 3 * 8 + 4
28
>>> 3 * (8 + 4)
36
>>>
```

In programming, a "variable" is a container in which a data value can be stored within the computer's memory. The stored value can then be referenced using the variable's name. The programmer can choose any name for a variable, except the Python keywords listed on the inside front cover of this book, and it is good practice to choose meaningful names that reflect the variable's content.

In Python's interactive mode, a variable is created by stating its name at the prompt, followed by an = and the value to be stored. The value may be numeric or may be a text "string" enclosed within single or double quote marks.

Variable names can be used in expressions, and the value they contain will be substituted in calculations or can be printed out using Python's **print()** function. The last printed expression gets automatically assigned to a special _ (underscore) variable that can be used in a subsequent calculation:

Expressions that produce fractional results are given as floating-point numbers – the fractions do not get lost.

 4 Create a variable containing a text string, then print out its stored value

 5 Next, create two variables containing numeric values, then multiply their stored values, printing out the result

 6 Now, add to the last printed number, printing the result

7 Finally, use the Python **round()** function to round the last printed number to two decimal places, printing the result

Python's arithmetic operators are...
+ (addition),
- (subtraction),
* (multiplication),
/ (division), and
% (modulus).

```
File  Edit  Tabs  Help
>>> topic = 'Python Price Calculator'
>>> print(topic)
Python Price Calculator
>>> tax = 12.5 / 100
>>> price = 100.50
>>> price * tax
12.5625
>>> price + _
113.0625
>>> round( _ , 2 )
113.06
>>>
```

Writing your first program

Python's interactive mode is useful as a simple calculator, but you can create programs for more extensive functionality. A Python program is simply a plain text file created with an editor, such as Leafpad or nano, which has been saved with a ".py" file extension. Python programs can be executed simply by stating the program file name after the **python** command at a terminal prompt.

Python programs can get input from the user with a **raw_input()** function and store that input in a variable for use by the program. This function can display a text string that has been specified within its parentheses and enclosed between quotes. The stored input value can be printed out by the **print()** function and may be concatenated (joined) with other text using the + operator.

Text strings may be enclosed in either single or double quotes, but not mixed – the start and end quotes must be of the same type.

hello.py

Launch a plain text editor, then on the first line type a statement to display a request message and assign user input to a variable when they hit Return
user = raw_input('I am Raspberry Pi. What is your name?')

Now, on the next line, type a statement to print out the user input concatenated within other text
print('Welcome ' + user + '\nHave a nice day!')

Then, save the program file as **hello.py** in your home (username) directory

The concatenated output here includes a **\n** newline character – so the response gets printed out on two lines.

Launch a terminal window in your home directory, then enter **python hello.py** and hit Return to execute the program – enter your name when asked

```
File  Edit  Tabs  Help
pi@raspberrypi:~ $ python hello.py
I am Raspberry Pi. What is your name? Mike
Welcome Mike
Have a nice day!
pi@raspberrypi:~ $ █
```

The program runs because the **python** command invokes the Python interpreter, but programs can be made directly executable at a prompt by making two simple changes:

- Add the interpreter's path at the start of the program file
- Give the program file executable permission

The interpreter's path must be added on the first line of the program after **#!** (shebang) characters and is typically located at **/usr/bin/python**. Alternatively, with Python, on the user path this can be stated independent of location as **#! /usr/bin/env python**.

Permission to be executable can be granted with the command **chmod 755** followed by the program's filename.

If you want to include a quote mark within a string it must be preceded by a \ backslash character to prevent the string being prematurely terminated.

5 Reopen the program file **hello.py** in a plain text editor and at the start of the very first line add the path **#! /usr/bin/env python**

6 Save the modified file, then at a prompt issue the command to make it executable by changing permissions **chmod 755 hello.py**

7 Finally, enter **./** (dotslash) followed by the program filename to execute the program directly **./hello.py**

```
File  Edit  Tabs  Help
pi@raspberrypi:~ $ cat hello.py
#!/usr/bin/env python
user = raw_input('I am Raspberry Pi. What is your name? ')
print('Welcome ' + user + '\nHave a nice day!')
pi@raspberrypi:~ $ chmod 755 hello.py
pi@raspberrypi:~ $ ls -l hello.py
-rwxr-xr-x 1 pi pi 129 Mar 11 07:21 hello.py
pi@raspberrypi:~ $ ./hello.py
I am Raspberry Pi. What is your name? Mike
Welcome Mike
Have a nice day!
pi@raspberrypi:~ $ █
```

The **cat** command is used here to confirm the program file has been modified, and the **ls -l** command is used to confirm the permissions have been changed.

Writing lists

In Python programming, a variable must be assigned an initial value (initialized) in the statement that declares it in a program, otherwise the interpreter will report a "not defined" error.

Multiple variables can be initialized with a common value in a single statement using a sequence of = assignments, like this:

a = b = c = 10   **# Initializes as a=10, b=10, c=10.**

Alternatively, multiple variables can be initialized with differing values in a single statement using comma separators, like this:

a , b , c = 1 , 2 , 3   **# Initializes as a=1 , b=2 , c=3.**

Unlike regular variables, which can only store a single item of data, a Python "list" is a variable that can store multiple items of data. The data is stored sequentially in list "elements" that are index numbered starting at zero. So, the first value is stored in element zero, the second value is stored in element one, and so on.

A list is created much like any other variable, but is initialized by assigning values as a comma-separated list between square brackets. For example, creating a list named "nums" like this:

nums = [0, 1, 2, 3, 4, 5]   **# Initializes a six element list.**

An individual list element can be referenced using the list name followed by square brackets containing that element's index number. This means that **nums[1]** references the second element in the example above – not the first element, as element numbering starts at zero.

Lists can have more than one index – to represent multiple dimensions, rather than the single dimension of a regular list. Multi-dimensional lists of three indices and more are uncommon, but two-dimensional lists are useful to store grid-based information such as X,Y coordinates.

A list of string values can even be considered to be a multi-dimensional list, as each string is itself a list of characters, so each character can be referenced by its index number within its particular string.

In Python, the **#** hash character is used to add comments to programs – anything that appears to the right of the **#** on a line is ignored by the Python interpreter.

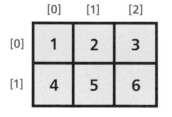

	[0]	[1]	[2]
[0]	1	2	3
[1]	4	5	6

1 Launch a plain text editor and begin a new Python program by locating the interpreter
#! /usr/bin/env python

list.py

2 Create a list of three elements containing string values
quarter = ['January', 'February', 'March']

3 Add statements to display the value in each list element
print('First Month: ' + quarter[0])
print('Second Month: ' + quarter[1])
print('Third Month: ' + quarter[2])

4 Create a multi-dimensional list of two elements, which themselves each contain three elements
coords = [['1', '2', '3'] , ['4', '5', '6']]

String indices may also be negative numbers – to start counting from the right where -1 references the last letter.

5 Add statements to display the value in two inner elements
print('Top Left 0,0: ' + coords[0][0])
print('Bottom Right Left 0,0: ' + coords[1][2])

6 Now, add a statement to display one character of a string
print('First letter in second month: ' + quarter[1][0])

7 Finally, save the file and make it executable with **chmod**, then run the program to see the list element values

```
File  Edit  Tabs  Help
pi@raspberrypi:~ $ chmod 755 list.py
pi@raspberrypi:~ $ ./list.py
First Month: January
Second Month: February
Third Month: March
Top Left 0,0: 1
Bottom Right 1,2: 6
First letter in second month: F
pi@raspberrypi:~ $ ▮
```

Loop structures, which are introduced later in this chapter, are often used to iterate through list elements.

Manipulating lists

List variables, which can contain multiple items of data, are widely used in Python programming and have a number of "methods" that can be dot-suffixed to the list name for manipulation:

For lists that contain both numerical and string values, the **sort()** method returns the list elements sorted first numerically then alphabetically – for example as 1,2,3,A,B,C.

List Method:	Description:
list.append(*x*)	Adds item *x* to the end of the list
list.extend(*L*)	Adds all items in list *L* to the list
list.insert(*i*,*x*)	Inserts item *x* at index position *i*
list.remove(*x*)	Removes first item *x* from the list
list.pop(*i*)	Removes item at index position *i* and returns it
list.index(*x*)	Returns the index position in the list of first item *x*
list.count(*x*)	Returns the number of times *x* appears in the list
list.sort()	Sort all list items, in place
list.reverse()	Reverse all list items, in place

Python also has a useful **len(L)** function that returns the length of the list **L** as the total number of elements it contains. Like the **index()** and **count()** methods, the returned value is numeric so cannot be directly concatenated to a text string for output.

Python also has an **int(s)** function that returns a numeric version of the string **s** value.

String representation of numeric values can, however, be produced by Python's **str(n)** function for concatenation to other strings, which returns a string version of the numeric **n** value. Similarly, a string representation of an entire list can be returned by the **str(L)** function for concatenation to other strings. In both cases, remember that the original version remains unchanged, as the returned versions are merely copies of the original version.

Individual list elements can be deleted by specifying their index number to the Python **del(i)** function. This can remove a single element at a specified **i** index position, or a "slice" of elements can be removed using slice notation **i1:i2** to specify the index number of the first and last element. In this case, **i1** is the index number of the first element to be removed and all elements up to, but not including, the element at the **i2** index number will be removed.

1 Launch a plain text editor and begin a new Python program by locating the interpreter
#! /usr/bin/env python

pop.py

2 Create two lists of three elements containing string values
basket = ['Apple', 'Banana', 'Cherry']
crate = ['Eggplant', 'Fig', 'Grape']

3 Add statements to display a list and its element count
print('Basket List: ' + str(basket))
print('Basket Elements: ' + str(len(basket)))

4 Next, add statements to add, then remove an element, displaying the list after each change
basket.append('Date')
print('Appended: ' + str(basket))
print('Last Item Removed: ' + basket.pop())
print('Basket List: ' + str(basket))

5 Now, add statements to extend and display a list
basket.extend(crate)
print('Extended: ' + str(basket))

6 Finally, add statements to delete a single element then a slice, displaying a list after each deletion
del basket[1]
print('Item Removed: ' + str(basket))
del basket[1:3]
print('Slice Removed: ' + str(basket))

7 Save the file and make it executable with **chmod**, then run the program to see the list get manipulated

The last index number in the slice denotes at what point to stop removing elements but the element at that position does not get removed.

```
File  Edit  Tabs  Help
pi@raspberrypi:~ $ chmod 755 pop.py
pi@raspberrypi:~ $ ./pop.py
Basket List: ['Apple', 'Banana', 'Cherry']
Basket Elements: 3
Appended: ['Apple', 'Banana', 'Cherry', 'Date']
Last Item Removed: Date
Basket List: ['Apple', 'Banana', 'Cherry']
Extended: ['Apple', 'Banana', 'Cherry', 'Eggplant', 'Fig', 'Grape']
Item Removed: ['Apple', 'Cherry', 'Eggplant', 'Fig', 'Grape']
Slice Removed: ['Apple', 'Fig', 'Grape']
pi@raspberrypi:~ $ ▮
```

Fixing in tuples

Lists and strings are similar, as a string is simply a list of characters. Individual elements in a list and individual characters in a string may both be addressed by their index position. Additionally, lists and strings can both be modified (are "mutable").

In contrast to lists and strings, a "tuple" is a comma-separated sequence of items that cannot be modified (is "immutable"). Although a tuple is apparently similar in nature to a list, they are mostly used for different purposes – list elements typically contain element values of similarity to each other, such as types of fruit, whereas tuple items are typically diverse and have no similarity.

This process is known as "tuple packing". The sequence of items stored inside a tuple can later be assigned to individual variables in a process known as "sequence unpacking".

The individual items within a tuple can be addressed by their sequence position, much as list elements can be referenced by their index position. Also, the Python **len()** function can be used to return the length of the sequence just as it can be used to return the length of a list.

Interestingly, a tuple, which cannot be modified after it has been assigned a sequence of items, may contain items that can themselves be modified. For instance, one of the items in the sequence assigned to a tuple might be a list. This means that individual elements of that list can be referenced using both the position number of the list in the tuple sequence and the index number of the element in the list.

A tuple can also contain other tuples nested in its sequence of items. When a tuple sequence is displayed in output it is surrounded by parentheses, so nested tuples are easily recognizable. Including parentheses in the statement creating a tuple is optional but could be considered good coding practice. The Python **str()** function can be used to concatenate a string representation of a tuple sequence to another string.

Like index numbering with lists, the items in a tuple sequence are numbered from zero.

For sequence unpacking, there must be the same number of variables as items in the sequence.

84

1 Launch a plain text editor and begin a new Python program by locating the interpreter
#! /usr/bin/env python

tuple.py

2 Create a list of three elements containing string values
zoo = ['Kangaroo', 'Leopard', 'Moose']

3 Next, create a tuple containing a sequence of three items, then display its entire contents
seq = (100, 'Bread', zoo)
print('Sequence: ' + str(seq))

4 Now, unpack the tuple sequence into individual variables
num, txt, list = seq

5 Then, add statements to display the unpacked items
print('Item 1: ' + str(num))
print('Item 2: ' + txt)
print('Item 3: ' + str(list))

6 Finally, add statements to modify one of the list element values within the tuple and display the modified sequence
seq[2][1] = 'Llama'
print('Modified: ' + str(seq))

7 Save the file and make it executable with **chmod**, then run the program to see the list modified inside the tuple

```
File  Edit  Tabs  Help
pi@raspberrypi:~ $ chmod 755 tuple.py
pi@raspberrypi:~ $ ./tuple.py
Sequence: (100, 'Bread', ['Kangaroo', 'Leopard', 'Moose'])
Item 1: 100
Item 2: Bread
Item 3: ['Kangaroo', 'Leopard', 'Moose']
Modified: (100, 'Bread', ['Kangaroo', 'Llama', 'Moose'])
pi@raspberrypi:~ $
```

Multiple assignment of values to variables, as described on page 80, is simply a combination of tuple packing and sequence unpacking.

Collecting in sets

In Python programming, a "set" is a data container, like lists and tuples, but a set can only contain an unordered collection of unique items – duplicates are not allowed. This means that sets are useful for eliminating duplicate entries and membership testing. Creating a set is simply a matter of assigning the items as a comma-separated list between curly brackets (braces) to a name of your choice. Sets have a number of "methods" that can be dot suffixed to the set name for manipulation and comparison:

Set Method:	Description:
set.add(*x*)	Adds item *x* to the set
set.update(*x,y,z*)	Adds multiple items to the set
set.copy()	Returns a copy of the set
set1.issubset(*set2*)	Returns True if *set1* is a subset of *set2*
set1.issuperset(*set2*)	Returns True if *set1* is a superset of *set2*
set.pop()	Removes one random item
set.discard(*i*)	Removes item at position *i*
set.clear()	Removes all items from the set
set1.union(*set2*)	Returns a united set of all unique items
set1.intersection(*set2*)	Returns items that are in both sets
set1.difference(*set2*)	Returns items that are in *set1* but are not in *set2*

A set can be searched to see if it contains a particular item with the Python **in** operator, using the syntax ***item* in *set***. The search will return a Boolean **True** value when the item is found, otherwise it will return **False**. The items in a set are not fixed and can be changed as the program proceeds (are "mutable") but if you wish to have fixed items that cannot be changed (are "immutable"), you can use the Python **frozenset()** constructor to freeze the set. A frozen set is like a regular set, except it is immutable, and can use the comparison methods above, but you cannot add items. The Python **str()** function can be used to concatenate a string representation of a set and a Boolean value to another string.

There is also a **remove(*i*)** set method but this reports an error if you try to remove a non-existent item – the **discard(*i*)** method performs the same task but never reports an error.

A set can also be created using the **set()** constructor function.

1 Launch a plain text editor and begin a new Python program by locating the interpreter
#! /usr/bin/env python

set.py

2 Create and display a set of three color items
bag = { 'Red', 'Green', 'Blue' }
print('Set: ' + str(bag))

3 Next, add an item to the set and display it once more
bag.add('Yellow')
print('Enlarged Set: ' + str(bag))

4 Now, search the set for an item and display the result
print('Is Green In Set?: ' + str('Green' in bag))

5 Then, create and display a frozen set of three color items
frozenbag = ({ 'Red', 'Purple', 'Yellow' })
print('Frozen Set: ' + str(frozenbag))

6 Finally, display only those items that appear in both sets
print('Common To Both Sets: ' +
       str(bag.intersection(frozenbag)))

7 Save the file and make it executable with **chmod**, then run the program to see the list modified inside the tuple

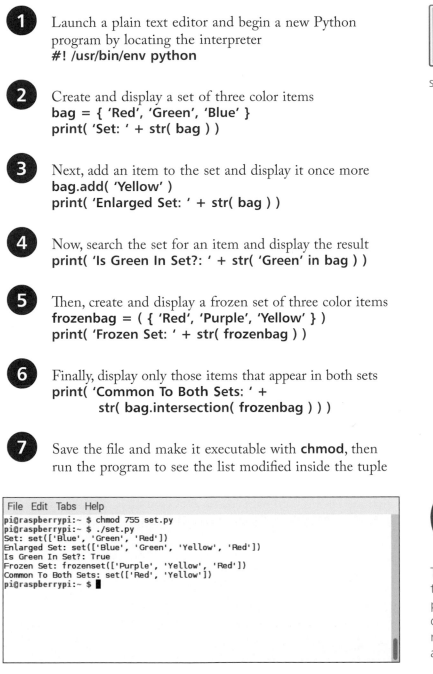

```
File Edit Tabs Help
pi@raspberrypi:~ $ chmod 755 set.py
pi@raspberrypi:~ $ ./set.py
Set: set(['Blue', 'Green', 'Red'])
Enlarged Set: set(['Blue', 'Green', 'Yellow', 'Red'])
Is Green In Set?: True
Frozen Set: frozenset(['Purple', 'Yellow', 'Red'])
Common To Both Sets: set(['Red', 'Yellow'])
pi@raspberrypi:~ $
```

Try adding an item to the frozen set to see the program report an error on execution – you are not allowed to change an immutable frozen set.

87

In other programming languages a list is often called an "array" and a dictionary is often called an "associative array".

Data is frequently associated as key:value pairs – for example, when you submit a web form, text value typed into an input field is typically associated with that text field's name as its key.

Associating in dictionaries

In Python programming, a "dictionary" is a data container that can store multiple items of data as a list of key:value pairs. Unlike list container values, which are referenced by their index number, values stored in dictionaries are referenced by their associated key. The key must be unique within that dictionary and is typically a string name, although numbers may be used.

Creating a dictionary is simply a matter of assigning the key:value pairs as a comma-separated list between curly brackets (braces) to a name of your choice. Strings must be enclosed within quotes, as usual, and a : colon character must come between the key and its associated value.

A key:value pair can be deleted from a dictionary by specifying the dictionary name and the pair's key to the **del** keyword. Conversely, a key:value pair can be added to a dictionary by assigning a value to the dictionary's name and a new key.

Python dictionaries have a **keys()** methods that can be dot suffixed to the dictionary name to return a list, in random order, of all the keys in that dictionary. If you prefer the keys to be sorted into alphanumeric order, simply enclose the statement within the parentheses of the Python **sorted()** function.

A dictionary can be searched to see if it contains a particular key with the Python **in** operator, using the syntax *key* **in** *dictionary*. The search will return a Boolean **True** value when the key is found in the specified dictionary, otherwise it will return **False**.

Dictionaries are the final type of data container available in Python programming. In summary, the various types are:

● **Variable** – stores a single value
● **List** – stores multiple values in an ordered index
● **Tuple** – stores multiple fixed values in a sequence
● **Set** – stores multiple unique values in an unordered collection
● **Dictionary** – stores multiple unordered key:value pairs

1 Launch a plain text editor and begin a new Python program by locating the interpreter
#! /usr/bin/env python

dict.py

2 Create a dictionary of three key:value pairs
**dict = { 'name' : 'Mike',
             'topic' : 'Python', 'system' : 'RasPi' }**

3 Next, display the entire dictionary
print('Dictionary: ' + str(dict))

4 Now, display a single value, referenced by its key
print('Topic: ' + str(dict['topic']))

5 Then, display all keys within the dictionary
print('Keys: ' + str(dict.keys()))

6 Delete one pair from the dictionary, then add another pair
del dict['name']
dict['user'] = 'Anne'
print('Dictionary: ' + str(dict))

7 Search the dictionary for a key and display the result
print('Is There A \'name\' Key?: ' + str('name' in dict))

8 Save the file and make it executable with **chmod**, then run the program to see the dictionary get modified

Notice that quotes must be preceded by a backslash escape character within a string – to prevent the string being prematurely terminated.

```
File  Edit  Tabs  Help
pi@raspberrypi:~ $ chmod 755 dict.py
pi@raspberrypi:~ $ ./dict.py
Dictionary: {'topic': 'Python', 'name': 'Mike', 'system': 'RasPi'}
Topic: Python
Keys: ['topic', 'name', 'system']
Dictionary: {'topic': 'Python', 'user': 'Anne', 'system': 'RasPi'}
Is There A 'name' Key?: False
pi@raspberrypi:~ $ 
```

Testing expressions

The Python **if** keyword performs the basic conditional test that evaluates a given expression for a Boolean value of **True** or **False**. This allows a program to proceed in different directions according to the result of the test and is known as "conditional branching". The tested expression must be followed by a : colon, then statements to be executed when the test succeeds should follow below on separate lines, and each line must be indented from the **if** test line. The size of the indentation is not important but it must be the same for each line.

Optionally, an **if** test can offer alternative statements to execute when the test fails by appending an **else** keyword after the statements to be executed when the test succeeds. The **else** keyword must be followed by a : colon and aligned with the **if** keyword, but its statements must be indented in a likewise manner.

An **if** test block can be followed by an alternative test using the **elif** keyword ("else if") that offers statements to be executed when the alternative test succeeds. This, too, must be aligned with the **if** keyword, followed by a : colon, and its statements indented.

The syntax for the **if**-**elif**-**else** structure looks like this:

if *test-expression-1* :
    statements-to-execute-when-test-expression-1-is-True
elif *test-expression-2* :
    statements-to-execute-when-test-expression-2-is-True
else :
    statements-to-execute-when-test-expressions-are-False

Typically, the test expression might make a comparison between two values using one of these Python comparison operators:

Indentation of code is very important in Python as it identifies code blocks to the interpreter – other programming languages use braces.

The **if: elif: else:** sequence is the Python equivalent of the **switch** or **case** statements found in other languages.

Operator:	Description:
==	Equality – returns True when values are equal
!=	Inequality – returns True when values are unequal
>	Greater Than, returns True when first value is higher than the second value
<	Less Than, returns True when first value is lower
>=	Greater Than Or Equal
<=	Less Than Or Equal

1 Launch a plain text editor and begin a new Python program by locating the interpreter
#! /usr/bin/env python

if.py

2 Request the user enters a number to initialize a variable
num = input('Please Enter A Number: ')

3 Next, test the variable and display an appropriate response
if num > 5 :
        print('Number Exceeds 5')
elif num < 5 :
        print('Number Is Less Than 5')
else :
        print('Number Is 5')

4 Now, test the variable again using two expressions and display an appropriate response only upon success
if num > 7 and num < 9 :
        print('Number Is 8')
if num == 1 or num == 3 :
        print('Number Is 1 Or 3')

Hot tip

The **and** keyword ensures the evaluation is **True** only when both tests succeed, whereas the **or** keyword ensures the evaluation is **True** when either test succeeds.

5 Save the file and make it executable with **chmod**, then run the program and enter a number to see the responses

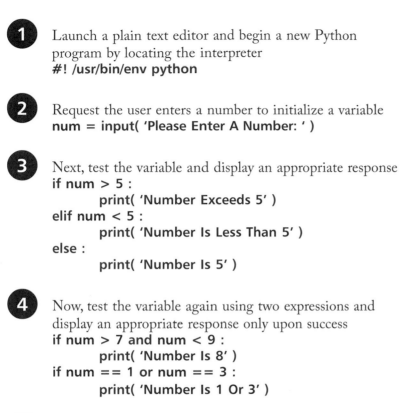

```
File Edit Tabs Help
pi@raspberrypi:~ $ chmod 755 if.py
pi@raspberrypi:~ $ ./if.py
Please Enter A Number: 4
Number Is Less Than 5
pi@raspberrypi:~ $ ./if.py
Please Enter A Number: 6
Number Exceeds 5
pi@raspberrypi:~ $ ./if.py
Please Enter A Number: 5
Number Is 5
pi@raspberrypi:~ $ ./if.py
Please Enter A Number: 8
Number Exceeds 5
Number Is 8
pi@raspberrypi:~ $ ./if.py
Please Enter A Number: 3
Number Is Less Than 5
Number Is 1 Or 3
pi@raspberrypi:~ $ 
```

Looping while true

A loop is a piece of code in a program that automatically repeats. One complete execution of all statements within a loop is called an "iteration" or "pass". The length of the loop is controlled by a conditional test made within the loop. While the tested expression is found to be **True**, the loop will continue – until the tested expression is found to be **False**, at which point the loop ends.

In Python programming, the **while** keyword creates a loop. It is followed by the test expression then a : colon character. Statements to be executed when the test succeeds should follow below on separate lines and each line must be indented by an equivalent space from the **while** test line. This statement block must include a statement that will at some point change the result of the test expression evaluation – otherwise an infinite loop is created.

Indentation of code blocks must also be observed in Python's interactive mode – like this example that produces a Fibonacci sequence of numbers from a **while** loop:

```
File  Edit  Tabs  Help
pi@raspberrypi:~ $ python
Python 2.7.9 (default, Mar  8 2015, 00:52:26)
[GCC 4.9.2] on linux2
Type "help", "copyright", "credits" or "license" for more information.
>>> a , b = 0 , 1
>>> while b < 100 :
...     print(b)
...     a , b = b , a+b
...
1
1
2
3
5
8
13
21
34
55
89
>>> quit()
pi@raspberrypi:~ $
```

Loops can be nested, one within another, to allow complete execution of all iterations of an inner nested loop on each iteration of the outer loop. A "counter" variable can be initialized with a starting value immediately before each loop definition, included in the test expression, and incremented on each iteration until the test fails – at which point the loop ends.

Unlike other Python keywords, the keywords **True** and **False** begin with uppercase letters.

The interpreter provides a ... continuation prompt when it expects further statements. Hit Tab to indent each statement, then hit Return to continue. Hit Return directly at the continuation prompt to discontinue.

1 Launch a plain text editor and begin a new Python program by locating the interpreter
#! /usr/bin/env python

while.py

2 Initialize a "counter" variable and define an outer loop using that variable in the test expression
i = 1
while i < 4 :

3 Next, add <u>indented</u> statements to display the counter's value and increment its value on each iteration of the loop
    print('Outer Loop Iteration ' + str(i))
    i += 1

The **+=** assignment statement **i += 1** is simply a shorthand way to say **i = i+1** – you can also use ***= /= -=** shorthand to assign values to variables.

4 Now, initialize a second "counter" variable and define an inner loop using this variable in the test expression, using statements indented to the same level as the previous step
    j = 1
    while j < 4 :

5 Finally, add <u>further-indented</u> statements to display this counter's value and increment its value on each iteration
    print('\tInner Loop Iteration ' + str(j))
    j += 1

The output printed from the inner loop is indented from that of the outer loop by the **\t** tab character.

6 Save the file and make it executable with **chmod**, then run the program to see the counter values on each pass

```
File  Edit  Tabs  Help
pi@raspberrypi:~ $ chmod 755 while.py
pi@raspberrypi:~ $ ./while.py
Outer Loop Iteration 1
        Inner Loop Iteration 1
        Inner Loop Iteration 2
        Inner Loop Iteration 3
Outer Loop Iteration 2
        Inner Loop Iteration 1
        Inner Loop Iteration 2
        Inner Loop Iteration 3
Outer Loop Iteration 3
        Inner Loop Iteration 1
        Inner Loop Iteration 2
        Inner Loop Iteration 3
pi@raspberrypi:~ $ 
```

Looping over items

In Python programming, the **for** keyword loops over all items in any list specified to the **in** keyword. This statement must end with a : colon character and statements to be executed on each iteration of the loop must be indented below, like this:

for *each-item* **in** *list-name* :
        statement-to-execute-on-each-iteration
        statement-to-execute-on-each-iteration

As a string is simply a list of characters, the **for in** statement can loop over each character. Similarly, a **for in** statement can loop over each element in a list, each item in a tuple, each member of a set, or each key in a dictionary:

for.py

The **for** loop in Python is unlike that in other languages such as C, as it does not allow step size and end to be specified.

 Launch a plain text editor and begin a new Python program by locating the interpreter
#! /usr/bin/env python

 Initialize a regular variable with a string value
user = 'Anne'

 Next, add a loop to display each character in the variable
for char in user :
        print(' String Character: ' + char)

4 Now, add a statement to print a separator line of asterisks
print('*' * 20)

 Then, create a list of three elements and a loop to display the value in each element
quarter = ['January', 'February', 'March']
for month in quarter :
        print('List Element: ' + month)

6 Add a statement to print another separator line of asterisks
print('*' * 20)

7 Next, create a tuple of three items and a loop to display each item, then print a separator line

```
seq = ( 100, 'Bread', quarter )
for item in seq :
        print( 'Tuple Item: ' + item )
print( '*' * 20 )
```

8 Now, create a set with three members and a loop to display each member, then print a separator line

```
bag = { 'Red', 'Green', 'Blue' }
for color in bag :
        print( 'Set Member: ' + color )
print( '*' * 20 )
```

9 Finally, create a dictionary with three pairs, then a loop to display each key and its associated value

```
dict = { 'name' : 'Mike',
                'topic' : 'Python', 'system' : 'Raspi' }
for key in dict :
        print( 'Dictionary Pair: ' + key
                            + ':' + dict[ key ] )
```

10 Save the file and make it executable with **chmod**, then run the program to see the values displayed by each loop

```
File  Edit  Tabs  Help
pi@raspberrypi:~ $ chmod 755 for.py
pi@raspberrypi:~ $ ./for.py
String Character: A
String Character: n
String Character: n
String Character: e
********************
List Element: January
List Element: February
List Element: March
********************
Tuple Item: 100
Tuple Item: Bread
Tuple Item: ['January', 'February', 'March']
********************
Set Member: Blue
Set Member: Green
Set Member: Red
********************
Dictionary Pair: topic:Python
Dictionary Pair: name:Mike
Dictionary Pair: system:RasPi
pi@raspberrypi:~ $ █
```

List elements are enclosed within [ ] square brackets, and tuple items within ( ) parentheses, but sets and dictionaries both use { } curly brackets.

Looping for a number

A **for** loop iterates over the items of any list or string in the order that they appear in the sequence, but you cannot directly specify the number of iterations to make, a halting condition, or the size of iteration step. You can, however, use the Python **range()** function to iterate over a sequence of numbers by specifying an numeric end value within its parameters. This will generate a sequence that starts at zero and continues up to, but not including, the specified end value. For example **range(5)** generates 0,1,2,3,4.

Optionally, you can specify both a start and end value within the parentheses of the **range()** function, separated by a comma. Also, you can specify a start value, end value, and a step value to the **range()** function as a comma-separated list within its parentheses.

To iterate over the elements in a list, you can specify the list size to the **range()** function using the **len()** function as the end value. Alternatively, you can specify the list's name within the parentheses of Python's **enumerate()** function to display each element's index number and value.

When looping through a dictionary you can display each key and its associated value using the dictionary **items()** method, and when looping through two lists simultaneously the elements can be paired together using Python's **zip()** function.

The **range()** function can generate a sequence that decreases, counting down, as well as those that count upward.

range.py

1 Launch a plain text editor and begin a new Python program by locating the interpreter
#! /usr/bin/env python

2 Create a loop of five iterations from a sequence that begins at one, ends at 20, and steps forward by fours – then display each step value and print a separator line
```
for i in range( 1, 20, 4 ) :
        print( 'Step: ' + str( i ) )
print( '*' * 20 )
```

3 Next, create a list of five elements then a loop to display each element value and print a separator line
```
langs = [ 'Python', 'Java', 'SQL', 'HTML', 'PHP' ]
for i in range( len( langs ) ) :
        print( 'Language: ' + langs[ i ] )
print( '*' * 20 )
```

4 Now, create a loop to display each element index number and value, then print a separator line

```
for i in enumerate( langs ) :
        print( 'Enumerated: ' + str( i ) )
print( '*' * 20 )
```

5 Then, create a dictionary of three pairs and a loop to display each key and value, then print a separator line

```
dict = { 'name' : 'Mike', 'topic' : 'Python', 'system' :
'RasPi' }
for i , j in dict.items() :
        print( 'Pair: ' + i + ':' + j )
print( '*' * 20 )
```

It is traditional to name "trivial" variables used as iteration counters simply with the letters **i**, **j**, **k**.

6 Finally, create another list of five elements then a loop to display each element value in both lists

```
chars = [ 'A', 'B', 'C', 'D', 'E' ]
for i , j in zip( chars, langs ) :
        print( 'Both: ' + i + ':' + j )
```

7 Save the file and make it executable with **chmod**, then run the program to see the values displayed by each loop

In Python programming, anything that contains multiple items that can be looped over is described as "iterable".

```
File  Edit  Tabs  Help
pi@raspberrypi:~ $ chmod 755 range.py
pi@raspberrypi:~ $ ./range.py
Step: 1
Step: 5
Step: 9
Step: 13
Step: 17
********************
Language: Python
Language: Java
Language: SQL
Language: HTML
Language: PHP
********************
Enumerated: (0, 'Python')
Enumerated: (1, 'Java')
Enumerated: (2, 'SQL')
Enumerated: (3, 'HTML')
Enumerated: (4, 'PHP')
********************
Pair: topic:Python
Pair: name:Mike
Pair: system:RasPi
********************
Both: A:Python
Both: B:Java
Both: C:SQL
Both: D:HTML
Both: E:PHP
pi@raspberrypi:~ $
```

Breaking out of loops

The Python **break** keyword can be used to prematurely terminate a loop when a specified condition is met. The **break** statement is situated inside the loop statement block and is preceded by a test expression. When the test returns **True**, the loop ends immediately and the program proceeds on to the next task. For example, in a nested inner loop it proceeds to the next iteration of the outer loop.

break.py

 1 Launch a plain text editor and begin a new Python program by locating the interpreter
#! /usr/bin/env python

 2 Create an outer loop that iterates three times
for i in range(1, 4) :

 3 Next, create an inner loop that also iterates three times
for j in range(1, 4) :

 4 Now, add a statement in the inner loop to display the counter number of each iteration
print('Running i=' + i + ' j=' + j)

5 Save the file and make it executable with **chmod**, then run the program to see the values displayed by each loop

Hot tip

Compare these nested for loops with the nested while loops example on page 92.

```
File  Edit  Tabs  Help
pi@raspberrypi:~ $ chmod 755 break.py
pi@raspberrypi:~ $ ./break.py
Running i=1 j=1
Running i=1 j=2
Running i=1 j=3
Running i=2 j=1
Running i=2 j=2
Running i=2 j=3
Running i=3 j=1
Running i=3 j=2
Running i=3 j=3
pi@raspberrypi:~ $ ▮
```

6 Now, insert this **break** statement at the very beginning of the inner loop block, to break out of the inner loop – then save the file and run the program once more

```
if i == 2 and j == 1 :
    print( 'Breaks inner loop at i=2 j=1' )
    break
```

```
File  Edit  Tabs  Help
pi@raspberrypi:~ $ ./break.py
Running i=1 j=1
Running i=1 j=2
Running i=1 j=3
Breaks inner loop at i=2 j=1
Running i=3 j=1
Running i=3 j=2
Running i=3 j=3
pi@raspberrypi:~ $ ▮
```

Here, the **break** statement halts all three iterations of the inner loop when the outer loop tries to run it the second time.

The Python **continue** keyword can be used to skip a single iteration of a loop when a specified condition is met. The **continue** statement is situated inside the loop statement block and is preceded by a test expression. When the test returns **True**, that one iteration ends and the program proceeds to the next iteration.

7 Insert this **continue** statement at the beginning of the inner loop block, to skip the first iteration of the inner loop – then save the file and run the program again

```
if i == 1 and j == 1 :
    print('Continues inner loop at i=1 j=1')
    continue
```

```
File  Edit  Tabs  Help
pi@raspberrypi:~ $ ./break.py
Continues inner loop at i=1 j=1
Running i=1 j=2
Running i=1 j=3
Breaks inner loop at i=2 j=1
Running i=3 j=1
Running i=3 j=2
Running i=3 j=3
pi@raspberrypi:~ $ ▮
```

Here, the **continue** statement just skips the first iteration of the inner loop when the outer loop tries to run it for the first time.

Defining functions

Previous examples in this chapter have used built-in functions of the Python programming language, such as the **print()** function. However, most Python programs contain a number of custom functions that can be called as required when the program runs.

A custom function is created using the **def** (definition) keyword followed by a name of your choice and **( )** parentheses. The programmer can choose any name for a function except the Python keywords listed on the inside front cover of this book, and the name of an existing built-in function. This must end with a : colon character, then the statements to be executed whenever the function gets called must appear on lines below and indented. Syntax of a function definition, therefore, looks like this:

```
def function-name ( ) :
        statement-to-be-executed
        statement-to-be-executed
        statement-to-be-executed
```

Once the function statements have been executed, program flow resumes at the point directly following the function call. This modularity is very useful in Python programming to isolate set routines so they can be called upon repeatedly.

Like Python's built-in **str()** function, which returns a string representation of a non-string value, many functions return a value when called. The **return** keyword returns a value that could be assigned to a variable for further processing by the program.

To create custom functions it is necessary to understand the accessibility ("scope") of variables in a program. Variables created outside functions can be referenced by statements inside functions – they have "global" scope. Conversely, variables created inside functions cannot be referenced from outside the function in which they have been created – these have "local" scope. The limited accessibility of local variables means that variables of the same name can be created in different functions without conflict.

If you want to make a local variable accessible elsewhere, it must first be declared with the Python **global** keyword followed by its name only. It may subsequently be assigned a value that can be referenced from anywhere in the program. Where a global variable and a local variable have the same name, the function will use the local version.

1 Launch a plain text editor and begin a new Python program by locating the interpreter
```
#! /usr/bin/env python
```

function.py

2 Request the user enter a number to initialize a variable
```
num = input( 'Please Enter A Number: ' )
```

3 Next, create a function to confirm the entry
```
def thanks() :
        print( 'Thanks For Entering ' + str( num ) )
```

4 Now, create a function to use the entry for arithmetic assigned to a local variable and return a string
```
def square() :
        result = num * num
        return str( num ) + ' Squared = ' + str( result )
```

5 Then, create a function to use the entry for more arithmetic assigned to a like-named local variable and return a result string
```
def cube() :
        result = num * num * num
        return str( num ) + ' Cubed = ' + str( result )
```

6 Finally, call each function to execute their statements
```
thank()
print( square() )
print( cube() )
```

A function can include a statement that calls another function, which may return a value for use in the calling function block.

7 Save the file and make it executable with **chmod**, then run the program and enter a number to call the functions

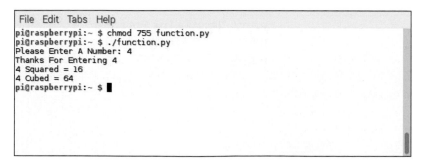

```
File Edit Tabs Help
pi@raspberrypi:~ $ chmod 755 function.py
pi@raspberrypi:~ $ ./function.py
Please Enter A Number: 4
Thanks For Entering 4
4 Squared = 16
4 Cubed = 64
pi@raspberrypi:~ $
```

101

Parameter naming follows the same conventions as variables and functions.

Supplying arguments

When defining a custom function in Python programming, you may optionally specify a parameter name between the function's parentheses. An "argument" value can then be passed to the function for use during its execution by referencing it via the parameter name. A function definition may allow multiple argument values to be passed to it by including a comma-separated list of parameter names within the function parentheses. When calling a function whose definition has multiple parameters, the call must pass the same number of argument values as the number of parameters specified in that function's definition. For example, **add(1, 2)** passes two integer arguments to **add(a, b)**.

It is important to consider data types in statements performing arithmetic or printing output. The Python **raw_input()** function returns a string data value that must be converted to an integer with **int()** to perform arithmetic, whereas an integer value must be converted to a string value with **str()** for concatenation to a string.

A string can be validated as numeric using its **isdigit()** method, then passed as an argument to other functions:

args.py

1 Launch a plain text editor and begin a new Python program by locating the interpreter
#! /usr/bin/env python

2 Create a main function that begins by requesting the user enter a number to initialize a variable
def main() :
        num = raw_input('Please Enter A Number':)

3 Next, in the main function, add statements that confirm the entry is a numeric string, then pass it as an argument to other functions
        if num.isdigit() :
                print('Thanks For Entering ' + num)
                print(num + ' Squared: ' + square(num))
                print(num + ' Cubed: ' + cube(num))

The **isdigit()** method only works for positive integers, so floating-point entries and negative integers are not valid.

4 Now, in the main function, add statements to display a message if the entry is non-numeric, then start over

```
else :
        print( 'Invalid Entry!' )
        main()
```

5 Create a function that requires a single argument to be passed to it from the caller

```
def square( num ) :
```

6 Next, add statements to the function that convert a passed string value to an integer to perform arithmetic, then return a result as a string value for concatenation

```
num = int( num )
return str( num * num )
```

7 Now, create another function that manipulates a passed argument value and returns a string result

```
def cube( num ) :
        num = int( num )
        return str( num * num * num )
```

8 Finally, add a call to the main function to start the program running

```
main()
```

9 Save the file and make it executable with **chmod**, then run the program and enter a non-numeric value to see the program restart or enter a number to pass to the functions

```
File  Edit  Tabs  Help
pi@raspberrypi:~ $ chmod 755 args.py
pi@raspberrypi:~ $ ./args.py
Please Enter A Number: MIKE
Invalid Entry!
Please Enter A Number: 7
Thanks For Entering 7
7 Squared = 49
7 Cubed = 343
pi@raspberrypi:~ $ ▮
```

Strictly speaking, a "parameter" (a.k.a. "formal parameter") is the name specified in a function definition, whereas an "argument" is the value passed in by a function call. You may, however, hear these terms used interchangeably.

103

Name parameters the same as variables passed to them to make the data movement obvious.

Summary

- The Python interpreter compiles program code and has an interactive mode that is started with the command **python**.
- A variable stores a data value that can be referenced using that variable's name and displayed using the **print()** function.
- Python programs are plain text files with a ".py" extension.
- A program can be made directly executable by stating the interpreter's path on the first line and setting permissions.
- Multiple assignments can be used to initialize several variables in a single statement.
- A list stores multiple items of data in sequentially numbered elements, starting at zero, which can be referenced using the list's name, followed by an index number in square brackets.
- Lists have several methods, such as **sort()**, that can be dot suffixed to a list's name for manipulation.
- The Python **len()** function returns the length of a specified list.
- A tuple is an immutable comma-separated sequence, a set is an unordered collection of unique items, and a dictionary is a comma-separated list of associated key:value pairs.
- The Python **if** keyword performs a conditional test on an expression for a Boolean value of **True** or **False**.
- Conditional branching provides alternatives with the **else** and **elif** keywords.
- A **while** loop repeats until a test expression returns **False**.
- A **for in** loop iterates over each item in a specified list or string.
- The **range()** function generates a numerical sequence that can be used to specify the length of a **for in** loop.
- The **break** and **continue** keywords interrupt loop iterations.
- A function is created using the **def** keyword and a name followed by parentheses.
- A function definition may optionally specify parameter names so that argument values may be passed in by function calls.

6 Importing modules

This chapter demonstrates how to use Python modules in your computer programs.

Storing functions

Python function definitions can usefully be stored in one or more separate files for easier maintenance, and to allow them to be used in several programs without copying the definitions into each one. Each file storing function definitions is called a "module" and the module name is the file name without the ".py" extension.

Functions stored in the module are made available to a program using the Python **import** keyword followed by the module name. Although not essential, it is customary to put any **import** statements at the beginning of the program. Imported functions can be called using their name dot-suffixed after the module name, e.g. a "steps" function from an imported module named "ineasy" can be called with **ineasy.steps()**.

Where functions stored in a module include parameters, it is often useful to assign a default argument value to the parameter in the definition. This makes the function more versatile, as it becomes optional for the call to specify an argument value:

A module need not include a line to locate the interpreter, as that will be included in the program file.

cat.py

 Launch a plain text editor, then begin a Python module with a function definition that supplies a default string value to its argument
```
def purr( pet = 'A cat' ) :
    print( pet + ' says MEOW!' )
```

 Next, add two more function definitions that also supply default string argument values to their parameters
```
def lick( pet = 'A cat' ) :
    print( pet + ' drinks milk' )

def nap( pet = 'A cat' ) :
    print( pet + ' sleeps by the fire' )
```

3 Now, save the file as "cat.py" so the module is named "cat"

There is no need to make the module file executable itself, as it is merely a library containing definitions that can be imported into an executable program. It should be placed in the same directory as the program files so the interpreter can easily find it. Optionally, you can create an alias when importing a module using **import as** keywords. For example, **import RPi.GPIO as GPIO** allows you to simply use **GPIO** as the function prefix in calls.

4 Launch a plain text editor and begin a new Python program by locating the interpreter
#! /usr/bin/env python

kitty.py

5 Add a statement to make the module functions available
import cat

6 Next, call each function without supplying an argument
cat.purr() ; cat.lick() ; cat.nap()

7 Now, call each function again and pass an argument to each, then save the file
cat.purr('Kitty') ; cat.lick('Kitty') ; cat.nap('Kitty')

8 Begin another program by locating the interpreter and making the module functions available once more
#! /usr/bin/env python
import cat

tiger.py

9 Request the user enters a name, then call each function passing the user-defined value as the argument
pet = raw_input('Enter A Pet Name: ')
cat.purr(pet) ; cat.lick(pet) ; cat.nap(pet)

10 Finally, save the file and make both files executable with **chmod**, then run the programs to see the functions called

```
File  Edit  Tabs  Help
pi@raspberrypi:~ $ chmod 755 kitty.py
pi@raspberrypi:~ $ chmod 755 tiger.py
pi@raspberrypi:~ $ ./kitty.py
A cat says MEOW!
A cat drinks milk
A cat sleeps by the fire
Kitty says MEOW!
Kitty drinks milk
Kitty sleeps by the fire
pi@raspberrypi:~ $ ./tiger.py
Enter A Pet Name: Tiger
Tiger says MEOW!
Tiger drinks milk
Tiger sleeps by the fire
pi@raspberrypi:~ $ ▮
```

Notice how multiple statements can appear on the same line if separated by a ; semi-colon character.

Owning function names

Internally, each Python module and program has its own "symbol table" which is used by all functions defined in that context only. This avoids possible conflicts with functions of the same name in another module if both modules were imported into one program.

When you import a module with an **import** statement, that module's symbol table does not get added to the program's symbol table – only the module's name gets added. That is why you need to call the module's functions using their module name prefix. Importing a "steps" function from a module named "ineasy" and another "steps" function from a module named "dance" lets them be called without conflict as **ineasy.steps()** and **dance.steps()**.

Generally, it is preferable to avoid conflicts by importing the module name and calling its functions with the module name prefix, but you can import individual function names instead with a **from import** statement. The module name is specified after the **from** keyword, and functions to import are specified as a comma-separated list after the **import** keyword. Alternatively, the * wildcard character can be specified after **import** to import all function names into the program's own symbol table. This means the functions can be called without a module name prefix:

Where you import individual function names, the module name does not get imported – so it cannot be used as a prefix.

dog.py

 1 Launch a plain text editor, then begin a Python module with a function definition that supplies a default string argument value to its parameter
```
def bark( pet = 'A dog' ) :
        print( pet + ' says WOOF!' )
```

 2 Next, add two more function definitions that also supply default string argument values to their parameters
```
def lick( pet = 'A dog' ) :
        print( pet + ' drinks water' )

def nap( pet = 'A dog' ) :
        print( pet + ' sleeps in the sun' )
```

3 Save the file as "dog.py" so the module is named "dog".

4 Launch a plain text editor and begin a new Python program by locating the interpreter
#! /usr/bin/env python

pooch.py

5 Add a statement to make the module functions available
from dog import bark , lick , nap

6 Next, call each function without supplying an argument
bark() ; lick() ; nap()

7 Now, call each function again and pass an argument value to each, then save the file
bark('Pooch') ; lick('Pooch') ; nap('Pooch')

8 Begin another program by locating the interpreter and making the module functions available once more
#! /usr/bin/env python
from dog import *

fido.py

9 Request the user enter a name, then call each function passing the user-defined value as the argument
pet = raw_input('Enter A Pet Name: ')
bark(pet) ; lick(pet) ; nap(pet)

10 Finally, save the file and make both files executable with **chmod**, then run the programs to see the functions called

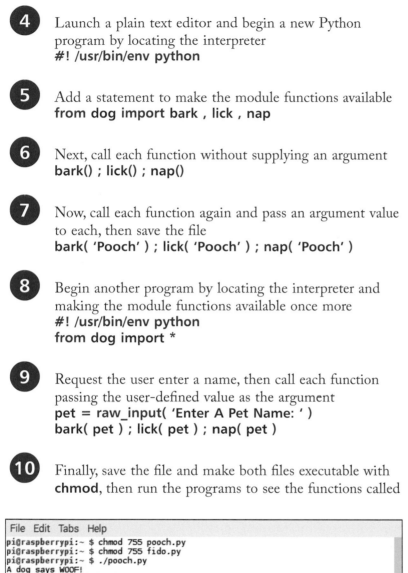

```
File Edit Tabs Help
pi@raspberrypi:~ $ chmod 755 pooch.py
pi@raspberrypi:~ $ chmod 755 fido.py
pi@raspberrypi:~ $ ./pooch.py
A dog says WOOF!
A dog drinks water
A dog sleeps in the sun
Pooch says WOOF!
Pooch drinks water
Pooch sleeps in the sun
pi@raspberrypi:~ $ ./fido.py
Enter A Pet Name: Fido
Fido says WOOF!
Fido drinks water
Fido sleeps in the sun
pi@raspberrypi:~ $ ▮
```

Hot tip

For larger programs you can import modules into other modules to build a module hierarchy.

You can find a description of all functions and methods in the documentation at docs.python.org/2.7

Formatting strings

The Python **dir()** function can be useful to examine the names of functions and variables defined in a module by specifying the module name within its parentheses. Interactive mode can easily be used for this purpose by importing the module name then calling the **dir()** function. The example below examines the "dog" module created on the previous page:

```
File  Edit  Tabs  Help
pi@raspberrypi:~ $ python
Python 2.7.9 (default, Mar  8 2015, 00:52:26)
[GCC 4.9.2] on linux2
Type "help", "copyright", "credits" or "license" for more information.
>>> import dog
>>> dir(dog)
['__builtins__', '__doc__', '__file__', '__name__', '__package__',
 'bark', 'lick', 'nap']
>>>
```

Those defined names that begin and end with a double underscore are Python objects, whereas the others are programmer-defined. The **__builtins__** module can also be examined using the **dir()** function, to examine the names of functions and variables defined by default, such as the **print()** function and a **str** object.

The **str** object defines several useful methods for string formatting, including an actual **format()** method that performs substitutions. The string to be formatted by this method can contain both text and "replacement fields" where the substitutions are to be made. Each replacement field is denoted by a pair of **{ }** braces containing a numerical value. That number is the index position of a replacement value in a comma-separated list specified within the parentheses of the **format()** method.

Other methods of the **str** object useful for string formatting include the **upper()** and **lower()** methods, which return a copy of the string with changed character case, the **capitalize()** method, which changes only the very first letter uppercase, and the **title()** method which makes only the first letter of each word uppercase.

Additionally, the case of each character can be reversed with the **swapcase()** method and space characters added, to pad the string with **rjust()** and **ljust()** methods.

The **str** object also has an **isalpha()** method, which returns **True** only when the string is entirely alphabetic, and an **isdigit()** method, which returns **True** only when all characters in the string are digits.

1 Launch a plain text editor and begin a new Python program by locating the interpreter
#! /usr/bin/env python

format.py

2 Next, add statements to initialize a variable with a formatted string and display the substitutions
snack = '{0} and {1}'.format('Burger' , 'Fries')
print('Substituted: ' + snack)

3 Now, add a statement to display the string with just the very first character in uppercase
print('Capitalized: ' + snack.capitalized())

4 Add statements to display the string entirely in uppercase then all first letters in uppercase
print('Uppercase: ' + snack.upper())
print('Titled: ' + snack.title())

5 Then, add a statement to display the string with uppercase characters swapped to lowercase and vice versa
print('Swapped: ' + snack.swapcase())

6 Finally, add a statement to display the string padded with spaces at the left to make its length total 20
print('Padded: ' + snack.rjust(20))

Do not confuse the **str** object described here with the **str()** function that converts values to the string data type.

7 Save the file and make it executable with **chmod**, then run the program to see the formatted strings

```
File  Edit  Tabs  Help
pi@raspberrypi:~ $ chmod 755 format.py
pi@raspberrypi:~ $ ./format.py
Substituted: Burger and Fries
Capitalized: Burger and fries
Uppercase: BURGER AND FRIES
Titled: Burger And Fries
Swapped: bURGER AND fRIES
Padded:     Burger and Fries
pi@raspberrypi:~ $ 
```

Reading & writing files

The **__builtins__** module can be examined using the **dir()** function to reveal that it contains a **file** object that has several methods for working with files, including **open()**, **read()**, **write()**, and **close()**.

Before a file can be read or written, it firstly must always be opened using the **open()** method. This requires two string arguments to specify the name and location of the file, and a "mode" in which to open the file:

If your program tries to open a non-existent file in **r** mode the interpreter will report an error.

File mode:	Operation:
r	Open an existing file to read.
w	Open an existing file to write. Creates a new file if none exists or opens an existing file and discards all its previous contents.
a	Appends text. Opens or creates a text file for writing at the end of the file.
r+	Opens a text file to read from or write to.
w+	Opens a text file to write to or read from.
a+	Opens or creates a text file to read from or write to at the end of the file.

Where the mode includes a **b** after any of the file modes listed above, the operation relates to a binary file rather than a text file. For example, **rb** or **w+b**.

Once a file has been successfully opened it can be read, or added to, or new text can be written in the file, depending on the mode specified in the call to the **open()** method. The open file must then always be closed by calling the **close()** method.

As you might expect, the **read()** method returns the entire content of the file and the **write()** method adds content to the file.

You can also use a **readlines()** method that returns a list of all lines.

You can quickly and efficiently read the entire contents in a loop, iterating line by line. The \n newline character marks the end of each line – but as the **print()** function adds another newline, output lines would be double-spaced. It is therefore preferable to use the **sys.stdout.write()** method, which does not add newlines, by importing the **sys** module to make it available.

1 Launch a plain text editor and begin a new Python program by locating the interpreter
#! /usr/bin/env python

file.py

2 Add a statement to make "sys" module functions available
import sys

3 Next, add a statement to initialize a variable with lines
poem = 'I never saw a man who looked\n'
poem += 'With such a wistful eye\n'
poem += 'Upon that little tent of blue\n'
poem += 'Which prisoners call the sky\n'

4 Now, add statements to write the lines into a text file
file = open('poem.txt' , 'w')
file.write(poem)
file.close()

5 Finally, add statements to read the lines from the text file and display them in output
file = open('poem.txt' , 'r')
for line in file :
        sys.stdout.write(line)
file.close()

The arguments to the **open()** method must be strings – enclosed between quote marks.

6 Save the file and make it executable with **chmod**, then run the program to write then read the text file

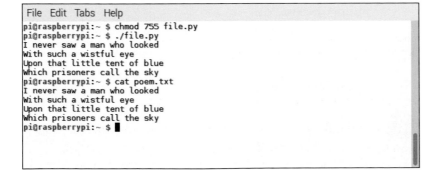

Pickling objects

In Python, string data can easily be stored in a text file using the technique demonstrated in the previous example. Other data types, such as numbers, lists, or dictionaries, could also be stored in text files but would require conversion to strings first. Restoring that stored data to their original data type on retrieval would require another conversion. An easier way to achieve data persistence of non-string data is provided by the "pickle" module.

The process of "pickling" objects stores a string representation of an object that can later be "unpickled" to its former state, and is a very common Python programming procedure.

An object can be converted for storage in a file by specifying the object and file as arguments to the **pickle** object's **dump()** method. It can be restored from that file by specifying the file name as the sole argument to the **pickle** object's **load()** method.

Unless the storage file needs to be human-readable for some reason, it is more efficient to use a machine-readable binary file.

Where the program needs to check for the existence of a storage file, the "os" module provides a **path** object with an **isfile()** method that returns **True** if a file specified within its parentheses is found:

data.py

1 Launch a plain text editor and begin a new Python program by locating the interpreter
#! /usr/bin/env python

2 Next, add a statement to make "pickle" and "os" module methods available
import pickle , os

3 Now, add a statement to test that a particular data file does not already exist
if not os.path.isfile('pickle.dat') :

4 If the file is not found, create a list of two elements
data = [0 , 1]

5 Then, request user data to be assigned to the list elements
data[0] = raw_input('Enter Your Name: ')
data[1] = raw_input('Enter Topic: ')

...cont'd

6 Next, add a statement to create a binary file for writing to
```
file = open( 'pickle.dat' , 'wb' )
```

7 Now, add statements to dump data into the binary file
```
pickle.dump( data , file )
```

8 After writing the file, remember to close it
```
file.close()
```

Pickling is the standard way to create Python objects that can be used in other programs.

9 Next, add alternative statements to open an existing file if found for reading from
```
else :
    file = open( 'pickle.dat' , 'rb' )
```

10 Now, add statements to load the stored data into a variable
```
data = pickle.load( file )
```

11 After reading the file, remember to close it
```
file.close()
```

12 Finally, add a statement to display the restored data
```
print( 'Welcome Back To ' + data[1]
                + ',' + data[0] )
```

13 Save the file and make it executable with **chmod**, then run the program, enter data and see that data get restored

```
File  Edit  Tabs  Help
pi@raspberrypi:~ $ chmod 755 data.py
pi@raspberrypi:~ $ ./data.py
Enter Your Name: Mike
Enter Topic: Python
pi@raspberrypi:~ $ ./data.py
Welcome Back To Python , Mike
pi@raspberrypi:~ $ ▮
```

Although this example just stores two string values in a list, pickling can store almost any type of Python object.

Handling exceptions

In Python programming there are two types of error that can occur – syntax errors and exceptions. A syntax error occurs when the interpreter encounters code that does not conform to the Python language rules, such as a missing colon, unmatched parentheses, misspelt function names, or incorrect indentation. Helpfully, the interpreter reports the error, indicating its nature:

```
File  Edit  Tabs  Help
pi@raspberrypi:~ $ python
Python 2.7.9 (default, Mar  8 2015, 00:52:26)
[GCC 4.9.2] on linux2
Type "help", "copyright", "credits" or "license" for more information.
>>> print('Hello World)
  File "<stdin>", line 1
    print('Hello World)
                       ^
SyntaxError: EOL while scanning string literal
>>>
```

Notice that the ^ points to the spot where the interpreter found a syntax error – a missing quote mark in this case.

An exception is an error that occurs during execution of a program, and once again the interpreter indicates its nature:

```
File  Edit  Tabs  Help
pi@raspberrypi:~ $ python
Python 2.7.9 (default, Mar  8 2015, 00:52:26)
[GCC 4.9.2] on linux2
Type "help", "copyright", "credits" or "license" for more information.
>>> num = int( raw_input('Enter a Number: ') )
Enter a Number: FIVE
Traceback (most recent call last):
  File "<stdin>", line 1, in <module>
ValueError: invalid literal for int() with base 10: 'FIVE'
>>>
```

The example above has correct syntax but expects the user to enter an integer numerically so a "ValueError" exception type occurs.

Python identifies many different exception types and gives them standard names that can be used to handle those exceptions in a **try except** statement. The **try** clause suppresses the default interpreter error report and allows the programmer to display a user-friendly message, specified in the **except** clause. Optionally, the interpreter's error message can be assigned to a variable using the **as** keyword, then be incorporated in the displayed message.

You can discover all the Python exception type names online at docs.python.org/2/ library/ exceptions.html

1 Launch a plain text editor and begin a new Python program by locating the interpreter
#! /usr/bin/env python

try.py

2 Start a "try" clause with the keyword and a colon
try :

3 Request the user enter a number to initialize a variable
num = int(raw_input('Enter A Number: '))

4 Next, add a statement to confirm the entry
print('Thanks For Entering ' + num)

5 Start an "except" clause with the keyword, exception type name, and assign the description – followed by a colon
except ValueError as description :

6 Now, add statements to display a user-friendly message and a description of the exception that has occurred
print('Oops! Looks Like An Invalid Entry: ')
print(str(description))

7 Save the file and make it executable with **chmod**, then run the program, enter an integer to see a confirmation, then enter a string to see the exception handled

```
File Edit Tabs Help
pi@raspberrypi:~ $ chmod 755 try.py
pi@raspberrypi:~ $ ./try.py
Enter A Number: 5
Thanks For Entering 5
pi@raspberrypi:~ $ ./try.py
Enter A Number: FIVE
Oops! Looks Like An Invalid Entry:
invalid literal for int() with base 10: 'FIVE'
pi@raspberrypi:~ $ 
```

An exception type name must be correctly capitalized – or a **NameError** will occur.

A Python class is an "object" with methods and attributes, and is the cornerstone of Object Oriented Programming (OOP).

It is customary to begin class names with an uppercase character.

Defining classes

A class is a data structure that can contain both variables and functions in a single entity. In Python, class variables are known as "attributes" and class functions are known as "methods".

The class definition, which specifies its attributes and methods, is a blueprint from which working copies ("instances") can be made.

All properties of a class definition are referenced by the word **self** – so an attribute named "sound" is **self.sound**. Additionally, all methods in a class definition must have **self** as their first argument – so a method named "talk" is **talk(self)**.

When a class instance is created, a special **__init__(self)** method is automatically called. Subsequent arguments can be added in its parentheses if values are to be passed to initialize attributes.

A class definition begins with the **class** keyword followed by a name and : colon character. Class methods are then defined like functions using the **def** keyword and must be indented. So the syntax of a Python class definition might look like this:

class *Class-name* **:**

> **def __init__(self ,** *value* **,** *value* **) :**
> > **self.***attribute-name* **=** *value*
> > **self.***attribute-name* **=** *value*
>
> **def** *method-name***(self) :**
> > **return self.***attribute-name*

An instance copy of a class is created by calling the class name and supplying its required number of arguments in its parentheses. This call returns the instance for assignment to an instance name. The syntax to create an instance of the class above looks like this:

instance-name **=** *Class-name***(** *value* **,** *value* **)**

Instance names can reference their class attributes and methods by dot-suffixing their names, e.g. *instance-name.attribute-name* and *instance-name.method-name()*.

Class definitions can be made in a module file that can be imported into a program, to make that data structure available, where instances of the class can be created for use in the program.

1 Launch a plain text editor and begin a new "Bird" class definition in a module file
class Bird :

super.py

2 Next, define the class initializer to expect a "sound" argument value, for assignment to a class attribute
def __init__(self , sound) :
        self.sound = sound

3 Now, define a class method to return the value of the class attribute when called
def talk(self) :
        return self.sound

4 Save the module file, then begin a new Python program by locating the interpreter
#! /usr/bin/env python

instance.py

5 Import the class from the module file to make its data structure available
from super import Bird

6 Next, create an instance of the class and specify a sound value argument
polly = Bird('Tweet,tweet!')

7 Now, call the class instance method to display in output
print('Polly says : ' + polly.talk())

Polly – an instance of the Bird class.

8 Save the program file and make it executable with **chmod**, then run the program to see output from a class instance

```
File Edit Tabs Help
pi@raspberrypi:~ $ chmod 755 instance.py
pi@raspberrypi:~ $ ./instance.py
Polly says: Tweet,tweet!
pi@raspberrypi:~ $ █
```

Inheriting features

In Python programming, a class can be created as a brand new class, like the "Bird" class in the previous example, or can be derived from an existing class. A derived ("sub") class inherits the properties of the parent ("super") class from which it is derived.

The ability to inherit attributes and methods from another class enables derived classes to use existing methods and attributes, redefine attribute values, and add further methods and attributes.

Derived class definitions can be made in a module file that can be imported into a program, to make their data structures available, where instances of those derived classes can be created for use in the program:

sub.py

Here, both derived classes explicitly specify a fixed **sound** attribute value but use the inherited **talk()** method to display its value – the second derived class adds an attribute and method.

1 Launch a plain text editor and import the class from the previous module file – to make its data structure available
from super import Bird

2 Next, begin a class derived from the imported class
class Dove(Bird) :

3 Now, redefine this class initializer and initialize a "sound" class attribute directly in the class definition instead
def __init__(self) :
self.sound = 'Coo,coo!\n'

4 Begin another class derived from the imported class
class Chicken(Bird) :

5 Next, redefine the class initializer and initialize "sound" and "flight" attributes directly in the class definition
def __init(self) :
self.sound = 'Cluck,cluck!'
self.flight = 'I\'m just a chicken
- I can\'t fly!'

6 Now, define an additional class method to return the value of the second class attribute when called
def fly(self) :
return self.flight

7 Save the new module file, then begin a Python program by locating the interpreter
#! /usr/bin/env python

derived.py

8 Import the derived classes from their module file to make their data structures available
from sub import Dove , Chicken

9 Next, create an instance of a derived class
joey = Dove()

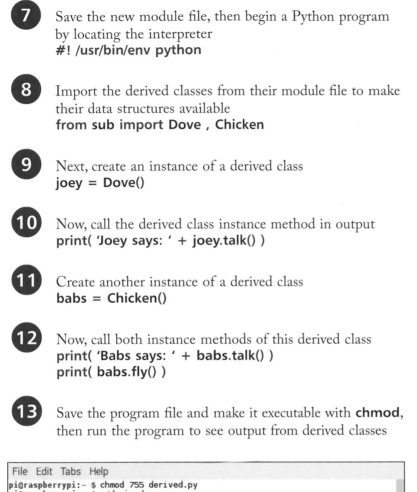

10 Now, call the derived class instance method in output
print('Joey says: ' + joey.talk())

11 Create another instance of a derived class
babs = Chicken()

Joey and Babs – both derived from Bird class.

12 Now, call both instance methods of this derived class
print('Babs says: ' + babs.talk())
print(babs.fly())

13 Save the program file and make it executable with **chmod**, then run the program to see output from derived classes

```
File  Edit  Tabs  Help
pi@raspberrypi:~ $ chmod 755 derived.py
pi@raspberrypi:~ $ ./derived.py
Joey says: Coo,coo!

Babs says: Cluck,cluck!
I'm just a chicken - I can't fly!
pi@raspberrypi:~ $ ▊
```

If you don't redefine the class initializer in derived classes, they will use that in the class from which they derive.

Interrogating the system

Python includes "sys" and "keyword" modules that are useful for interrogating the Python system itself. The "keyword" module contains a list of all Python keywords in its **kwlist** attribute and provides an **iskeyword()** method if you want to test a word.

You can explore the many features of the "sys" module, and indeed any feature of Python using the Interactive Mode help system. Just type **help()** at the **>>>** prompt to start the help system, then type **sys** at the **help>** prompt that appears.

Perhaps most usefully, the "sys" module has attributes that contain the Python version number, interpreter location on your system, and a list of all directories where the interpreter seeks module files – so if you save module files in any of these directories you can be sure the interpreter will find them:

system.py

 Launch a plain text editor and begin a new Python program by locating the interpreter
#! /usr/bin/env python

2 Next, import the "sys" and "keyword" modules to make their attributes and methods available
import sys , keyword

3 Now, add a statement to display the Python version
print('Python Version: ' + sys.version)

4 Add a statement to display the actual location on your Raspberry Pi of the Python interpreter
print('Python Interpreter Location: ' + sys.executable)

5 Now, add statements to display a list of all directories where the Python interpreter looks for module files
print('Python Module Search Path: ')
for dir in sys.path :
    print(dir)

6 Add statements to display a list all the Python keywords
print('Python Keywords: ')
for word in keyword.kwlist :
    print(word)

Save the file and make it executable with **chmod,** then run
the program to see the Python system information

```
File  Edit  Tabs  Help
pi@raspberrypi:~ $ chmod 755 system.py
pi@raspberrypi:~ $ ./system.py

Python Version: 2.7.9 (default, Mar  8 2015, 00:52:26)
[GCC 4.9.2]

Python Interpreter Location: /usr/bin/python

Python Module Search Path:
/home/pi
/usr/lib/python2.7
/usr/lib/python2.7/plat-arm-linux-gnueabihf
/usr/lib/python2.7/lib-tk
/usr/lib/python2.7/lib-old
/usr/lib/python2.7/lib-dynload
/usr/local/lib/python2.7/dist-packages
/usr/lib/python2.7/dist-packages
/usr/lib/python2.7/dist-packages/gtk-2.0
/usr/lib/pymodules/python2.7

Python Keywords:
and
as
assert
break
class
continue
def
del
elif
else
except
exec
finally
for
from
global
if
import
in
is
lambda
not
or
pass
print
raise
return
try
while
with
yield
pi@raspberrypi:~ $ ▮
```

The first item on the
Python search path is
your username directory
– so any file within
there, or within any
subdirectories you make
there, will be found by
the Python interpreter.

Spend a little time with
the Interactive Mode
help utility to discover
lots more about Python.

123

Performing mathematics

Python includes a "math" module that provides lots of methods you can use to perform mathematical procedures once imported.

The **math.ceil()** and **math.floor()** methods enable a program to perform rounding of a floating point value specified between their parentheses to the closest integer – **math.ceil()** rounds up and **math.floor()** rounds down but the value returned, although an integer, is a **float** data type rather than an **int** data type.

Integers can be cast from the **int** data type to the **float** data type using the **float()** function and to the **string** data type using the **str()** function.

The **math.pow()** method requires two arguments to raise a specified value by a specified power. The **math.sqrt()** method on the other hand simply requires a single argument, and returns the square root of that specified value. Both method results are returned as a numeric value of the **float** data type.

Typical trigonometry can be performed using methods from the math module, such as **math.sin()**, **math.cosin()** and **math.tan()**.

Additionally, Python includes a "random" module that can be used to produce pseudo random numbers once imported into a program.

The **random.random()** method produces a single floating-point number between zero and 1.0. Perhaps, more interestingly, the **random.sample()** method produces a list of elements selected at random from a sequence. This method requires two arguments to specify the sequence to select from, and the length of the list to be produced. As the **range()** function returns a sequence of numbers, this can be used to specify a sequence as the first argument to the **random.random()** method – so it will randomly select numbers from that sequence to produce a list in which no numbers repeat:

maths.py

1 Launch a plain text editor and begin a new Python program by locating the interpreter
#! /usr/bin/env python

2 Next, import the "math" and "random" modules to make their attributes and methods available
import math , random

3 Now, add statements to display two rounded values
print('Rounded Up 9.5 : ' + str(math.ceil(9.5)))
print('Rounded Down 9.5 : ' + str(math.floor(9.5)))

4 Add a statement to initialize a variable with an integer
num = 4

5 Next, add statements to display the square and square root of the variable value

```
print( str( num ) + ' Squared: '
                        + str( math.pow( num, 2 ) ) )
print( str( num ) + ' Square Root: '
                        + str( math.sqrt( num ) ) )
```

All the math methods here return floating-point numbers of the **float** data type.

6 Now, add a statement to produce a random list of six unique numbers between one and 59

nums = random.sample(range(1, 59) , 6)

7 Finally, add a statement to display the random list

print('Your Lucky Lotto Numbers: ' + str(nums))

8 Save the file and make it executable with **chmod**, then run the program several times to see the maths performed and to see the random sequence is unique each time

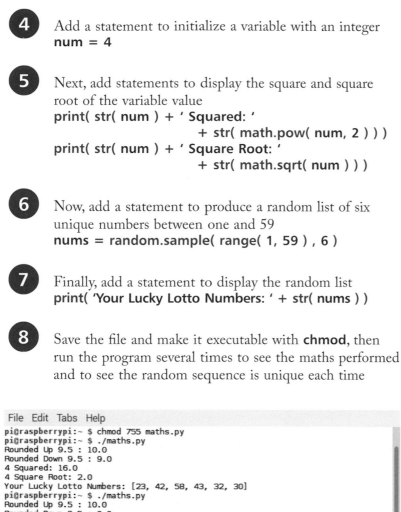

```
File Edit Tabs Help
pi@raspberrypi:~ $ chmod 755 maths.py
pi@raspberrypi:~ $ ./maths.py
Rounded Up 9.5 : 10.0
Rounded Down 9.5 : 9.0
4 Squared: 16.0
4 Square Root: 2.0
Your Lucky Lotto Numbers: [23, 42, 58, 43, 32, 30]
pi@raspberrypi:~ $ ./maths.py
Rounded Up 9.5 : 10.0
Rounded Down 9.5 : 9.0
4 Squared: 16.0
4 Square Root: 2.0
Your Lucky Lotto Numbers: [5, 56, 55, 19, 1, 31]
pi@raspberrypi:~ $ ./maths.py
Rounded Up 9.5 : 10.0
Rounded Down 9.5 : 9.0
4 Squared: 16.0
4 Square Root: 2.0
Your Lucky Lotto Numbers: [22, 18, 21, 19, 13, 44]
pi@raspberrypi:~ $ ./maths.py
Rounded Up 9.5 : 10.0
Rounded Down 9.5 : 9.0
4 Squared: 16.0
4 Square Root: 2.0
Your Lucky Lotto Numbers: [33, 13, 31, 24, 2, 42]
pi@raspberrypi:~ $ ./maths.py
Rounded Up 9.5 : 10.0
Rounded Down 9.5 : 9.0
4 Squared: 16.0
4 Square Root: 2.0
Your Lucky Lotto Numbers: [15, 14, 27, 1, 31, 11]
pi@raspberrypi:~ $ 
```

The list produced by **random.sample()** does not actually replace elements of the sequence but merely copies a sample, as its name says.

Calculating decimals

Python programs that attempt floating-point arithmetic can produce unexpected and inaccurate results, because the floating-point numbers cannot accurately represent all decimal numbers:

decimals.py

 1 Launch a plain text editor and begin a new Python program by locating the interpreter
#! /usr/bin/env python

 2 Next, initialize two variables with floating-point values
item = 0.70
rate = 1.05

 3 Now, initialize two more variables by attempting floating-point arithmetic with the first two variables
tax = item * rate
total = item + tax

Here, the **str.format()** method (introduced on page 110) includes a **:.2f** format specifier in the replacement field to return a string with two decimal places.

4 Add statements to display variable values, string formatted to have two decimal places so trailing zeros are shown
print('Item:\t' + '{:.2f}'.format(item))
print('Tax:\t' + '{:.2f}'.format(tax))
print('Total:\t' + '{:.2f}'.format(total))

5 Save the file and make it executable with **chmod**, then run the program to see the output of inaccurate addition

```
File  Edit  Tabs  Help
pi@raspberrypi:~ $ chmod 755 decimals.py
pi@raspberrypi:~ $ ./decimals.py
Item:    0.70
Tax:     0.73
Total:   1.44
pi@raspberrypi:~ $
```

 6 To help understand this problem, edit all three print statements to display the variable values string formatted to <u>20</u> decimal places, then run the program again
print('Item:\t' + '{:.20f}'.format(item))
print('Tax:\t' + '{:.20f}'.format(tax))
print('Total:\t' + '{:.20f}'.format(total))

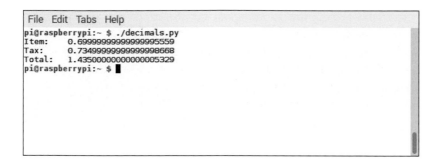

```
File Edit Tabs Help
pi@raspberrypi:~ $ ./decimals.py
Item:    0.69999999999999995559
Tax:     0.73499999999999998668
Total:   1.43500000000000005329
pi@raspberrypi:~ $ 
```

It is now clear that the tax value is represented numerically slightly below 0.735 so gets rounded down to 0.73. Conversely, the total value is represented numerically slightly above 1.435 so gets rounded up to 1.44, creating the apparent addition error.

Errors in floating-point arithmetic can be avoided using Python's "decimal" module. This provides a **Decimal()** object with which floating-point numbers can be more accurately represented:

7 Add a statement at the beginning of the program to import the "decimal" module to make all features available
from decimal import *

8 Next, edit the first two variable assignment to objects
item = Decimal(0.70)
rate = Decimal(1.05)

9 Save the changes then run the program again to see both tax and total representations will now get rounded down – so the output will show accurate addition when string formatting is changed back to two decimal places

```
File Edit Tabs Help
pi@raspberrypi:~ $ ./decimals.py
Item:    0.70
Tax:     0.73
Total:   1.43
pi@raspberrypi:~ $ 
```

Always use the **Decimal()** object to calculate monetary values or anywhere that accuracy is essential.

127

This problem is not unique to Python – Java has a **BigDecimal** class that overcomes this problem in much the same way as the **decimal** module found in Python.

Telling the time

The Python "datetime" module can be imported into a program to work with times and dates. It provides a **datetime** object with attributes of **year**, **month**, **day**, **hour**, **minute**, **second**, and **microsecond**. A **datetime** object has a **today()** method that assigns the current date and time values to its attributes and returns them in a tuple. It also has a **getattr()** method that requires two arguments specifying the datetime object name and attribute to retrieve. Alternatively, the attributes can be referenced using dot notation such as **datetime.year**.

All values in a **datetime** object are stored as numeric values but can be usefully transformed into text equivalents using its **strftime()** method. This requires a single string argument that is a "directive" specifying which part of the tuple to return, and in what format. The possible directives are listed in the table below:

As the datetime object is in a module of the same name, simply importing the module means it would be referenced as **datetime.datetime**. Use **from datetime import *** so it can be referenced just as **datetime**.

As the **strftime()** method requires a string argument, the specified directive must be enclosed between quote marks.

Directive:	Returns:
%A	Full weekday name (**%a** for abbreviated day)
%B	Full month name (**%b** for abbreviated month)
%c	Date and time appropriate for locale
%d	Day of the month number 1-31
%f	Microsecond number 0-999999
%H	Hour number 0-23 (24-hour clock)
%I	Hour number 1-12 (12-hour clock)
%j	Day of the year number 0-366
%m	Month number 1-12
%M	Minute number 0-59
%p	AM or PM equivalent for locale
%S	Second number 0-59
%w	Week day number 0(Sunday)-6
%W	Week of the year number 0-53
%X	Time appropriate for locale (**%x** for date)
%Y	Year 0001-9999 (**%y** for year 00-99)
%z	Timezone offset from UTC as +/-HHMM
%Z	Timezone name

1 Launch a plain text editor and begin a new Python program by locating the interpreter
#! /usr/bin/env python

time.py

2 Next, import the "datetime" module to make it available
from datetime import *

3 Now, create a datetime object with attributes assigned current values, then display its contents
today = datetime.today()
print('Today Is: ' + str(today))

4 Add a loop to display each attribute value individually
**for attr in **
['year' , 'month' , 'day' ,
**   'hour' , 'minute' , 'second' , 'microsecond'] :**
**   print(attr + ':\t' + str(getattr(today, attr)))**

Notice how the \ backslash character is used in this loop to allow a statement to continue on the next line without causing an error.

5 Next, add a statement to display time using dot notation
print('Time: ' + str(today.hour) + ':'
**   + str(today.minute))**

6 Now, assign formatted day and month names to variables
day = today.strftime('%A')
month = today.strftime('%B')

7 Finally, add a statement to display the formatted date
print('Date: '+ day +', '+ month +' '+ str(today.day))

8 Save the program file and make it executable with **chmod**, then run the program to see the time and date

```
File Edit Tabs Help
pi@raspberrypi:~ $ chmod 755 time.py
pi@raspberrypi:~ $ ./time.py
Today Is: 2016-03-12 10:53:04.387119
year:    2016
month:   3
day:     12
hour:    10
minute: 53
second: 4
microsecond:    387119
Time: 10:53
Date: Saturday, March 12
pi@raspberrypi:~ $
```

You can assign new values to attributes of a datetime object using its **replace()** method, such as **replace(year=2017)**.

Summary

- Python function definitions can usefully be stored in a module file then made available in a program by an **import** statement.
- An **import** statement only adds the module name to the program's symbol table, but **from import** adds function names.
- The **dir()** function lists the contents of a specified object.
- Strings can be formatted using methods of the **str** object such as **upper()**, **lower()**, **title()** and **format()**.
- Files can be created using methods of the **file** object that is readily available from the Python **__builtins__** module.
- Non-string data can be stored in, and retrieved from, binary files using the **dump()** and **load()** methods of the **pickle** object.
- Syntax errors occur when the interpreter encounters incorrect code, but exception errors occur during program execution.
- A **try except** statement can be used to handle exceptions.
- A class is a data structure of attributes and methods that is created using the **class** keyword, and references its properties internally using the **self** keyword.
- An instance is a working copy of a class.
- A derived sub class inherits the properties of its super class.
- The **sys** and **keyword** modules contain Python system details.
- Unlike the **print()** function, the **sys.stdout.write()** method displays output without adding a newline.
- The **math** module provides lots of methods to perform mathematical procedures, such as **ceil()**, **floor()**, **pow()** and **sqrt()**.
- Errors in floating-point arithmetic can be avoided by using the **Decimal()** object to more accurately represent float values.
- A **datetime** object contains date and time information as a tuple of **year**, **month**, **day**, **hour**, **minute**, **second**, and **microsecond**.

7 Producing games

This chapter demonstrates how to create graphical game components with the Python pygame module.

Creating a game window

Raspberry Pi ships with one or more games, such as Minecraft Pi for you to try out, but also includes the Python "pygame" module that you can use to make your own games.

The **pygame** module can be imported into a program like any other module to provide attributes and methods for games. Incorporated within this module is a "locals" module that contains useful game constants, such as **MOUSEBUTTONDOWN**. It is preferable to make these available to a program with a **from import** statement so they can be referenced without a prefix.

Every **pygame** program can usefully begin by calling its **init()** method to initialize all imported modules. Then the game window can be created by specifying a width and height pixel value to its **display.set_mode()** method. The window size will not be adjustable during the game, so its properties can be assigned to a program "constant" – like a variable, but with a fixed value.

The game window can be constantly refreshed by calling the **pygame.update()** method in a loop to redraw its surface. This loop can include statements to capture "events" that occur during execution of the program using the **pygame.event.get()** method.

Each event that occurs during execution of a **pygame** program can be recognized as being of a particular type, and statements can be provided to respond to each event. One event that must always be recognized is that which occurs when the user ends the game – typically by clicking the game window's "X" button to close it. This event is recognized by the **QUIT** constant that is provided by the **pygame.locals** module.

In response to the **QUIT** event, the **pygame.quit()** method should be called to terminate the resources started by **pygame.init()**. This does not completely exit the program, however. In order to shut down the Python interpreter connection, the "sys" module must be imported so its **quit()** method can also be called in response to the **QUIT** event.

The steps listed opposite create an empty game window that is the foundation for all other examples in this chapter to demonstrate various aspects of game development with the **pygame** module.

You can discover more about Pygame online at www.pygame.org

It is conventional to use uppercase for constant names to distinguish them from variables.

…cont'd

1 Begin a new Python program by locating the interpreter and making game methods and constants available
#! /usr/bin/env python
import pygame , sys
from pygame.locals import *

window.py

2 Next, add a statement to initialize this game on execution
pygame.init()

3 Now, define a window display size of 400 x 300 pixels
ZONE = pygame.display.set_mode((400, 300))

4 Then, add a game loop to continually update the display
while True :
        # Event capture statements go here...

        pygame.display.update()

5 Finally, insert statements into the game loop to capture the game events and stop the game when the user quits
            for event in pygame.event.get() :
                if event.type == QUIT :
                        pygame.quit()
                        sys.exit()

Clicking the "X" button on the game window creates a **QUIT** event so the game ends and the window closes.

6 Save the file and make it executable with **chmod**, then run the program to see an empty game window appear

```
File  Edit  Tabs  Help
pi@raspberrypi:~ $ chmod 755 window.py
pi@raspberrypi:~ $ ./window.py
```

pygame window

Painting shapes

By default, a game window displays the caption "pygame window" but you can specify a caption to a **pygame.display.set_caption()** method as a string argument within its parentheses.

Colors to paint on the window can be assigned to constants for easy reference as a list of three elements, specifying their red, green and blue components respectively. Each value must be in the range 0-255, where 255 is maximum for each color. For example, the list **(255, 0, 0)** assigns maximum red with no green or blue.

The entire game window ("surface") can be painted by specifying a color to its **fill()** method, then shapes painted onto it using methods of the **pygame.draw** object.

Rectangles are painted by the **pygame.draw.rect()** method. This requires three arguments to specify the surface constant name, a color, and a list of four numeric values – the X and Y coordinates of the rectangle's top left corner, its width, and its height.

Circles are painted by the **pygame.draw.circle()** method. This requires four arguments to specify the surface constant name, a color, the X and Y coordinates of the circle's center, and its radius.

Polygons are painted by the **pygame.draw.polygon()** method. This takes three arguments to specify the surface constant name, a color, and a series of X and Y coordinates describing its shape. The area between the first and last points will be filled with color.

Lines are painted by the **pygame.draw.line()** method. This requires five arguments to specify the surface constant name, a color, the x and y coordinates of a start point, the X and Y coordinates of an end point, and a width value.

Ellipses are painted by the **pygame.draw.ellipse()** method. This requires three arguments to specify the surface constant name, a color, and a list of four numeric values. Those describe the X and Y coordinates of a rectangle's top left corner, its width and height. The rectangle merely describes a boundary box inside which the ellipse will get painted.

Specifying colors as their proportions of red, green and blue is known as RGB notation.

...cont'd

1 Copy the empty game window program from page 133 then insert a statement to specify a caption
pygame.display.set_caption('Game Zone')

paint.py

2 Next, insert statements to define six colors
RED = (255, 0, 0) ; GREEN = (0, 255, 0)
BLUE = (0, 0, 255) ; YELLOW = (255, 255, 0)
WHITE = (255, 255, 255) ; FUCHSIA = (255, 0, 255)

3 Now, insert a statement to paint the entire game surface
ZONE.fill(BLUE)

4 Finally, insert statements to paint shapes on the surface
pygame.draw.rect(ZONE, RED, (25, 25, 100, 100))
pygame.draw.circle(ZONE, GREEN, (200,75), 50))
**pygame.draw.polygon(ZONE, YELLOW, **
**   (275,125), (325,25), (375,125)))**

**pygame.draw.line(ZONE, WHITE, **
**   (25,150), (375,150), 10))**

**pygame.draw.ellipse(ZONE, FUCHSIA, **
**   (25, 180, 350, 100))**

All statements in this example are inserted into the program code after **ZONE** has defined the window size but before the game loop.

135

5 Save the file and make it executable with **chmod**, then run the program to see shapes paint on the game window

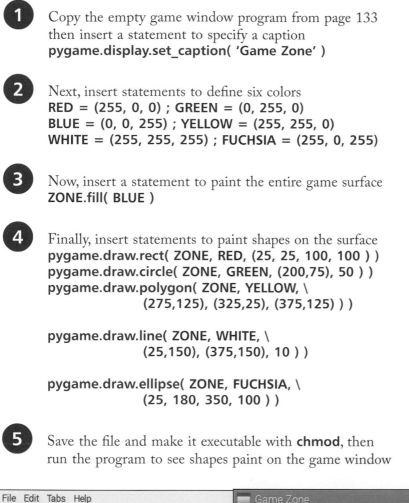

Blitting objects

Image objects can be created for display on a game window surface using the **pygame.image.load()** method. A source image file is specified between its parentheses and the returned image object assigned to a program constant. The source image can be in any one of these graphic file formats:

- **JPG** – Joint Photographic (experts) Group
- **PNG** – Portable Network Graphics
- **GIF** – Graphic Interchange Format (not animated)
- **BMP** – BitMaP
- **PCX** – Personal Computer eXchange
- **TGA** – TarGA
- **LBM** – interLeaved BitMap
- **PBM** – Portable BitMap
- **TIF** – Tagged Image File
- **XPM** – X-windows Pixel Map

The image object can then be copied onto the game surface using the surface's **blit()** method. This requires two arguments to specify the image object to copy, and the XY coordinates on the surface at which to position its top left corner.

Text can be written onto the game window surface in a similar way by first creating a **Font** object. This is created by specifying a font and size as arguments to the **pygame.font.Font()** constructor. The **Font** object has a **render()** method that returns a completed text block. This requires four arguments to specify the text string, whether to antialias the text (as a Boolean value of **True** or **False**), the color of the text, and the background color of the block.

The text block can then be copied onto the game surface using the surface's **blit()** method, specifying the text block to copy and the XY coordinates on the surface at which to position the top left corner of the block.

You can only render single lines of text – newline characters are not supported.

...cont'd

1 Copy the empty game window program from page 133 then insert a statement to specify a caption
pygame.display.set_caption('Game Zone')

blit.py

2 Next, insert statements to define three colors and to paint the entire game surface
RED = (255,0,0) ; YELLOW = (255,255,0) ;
BLUE = (0,0,255) ; ZONE.fill(BLUE)

3 Now, load an adjacent image into a constant
LOGO = pygame.image.load('raspi_logo.gif')

raspi_logo.gif

4 Then, add a statement to copy the image onto the surface
ZONE.blit(LOGO , (10, 10))

5 Next, add statements to define a font and text string
FONT = pygame.font.Font('freesansbold.ttf' , 32)
TEXT = FONT.render('Python Gaming',
                              True, RED, YELLOW)

freesansbold.ttf

6 Then, add a statement to copy the text onto the surface
ZONE.blit(TEXT , (100, 150))

7 Save the file and make it executable with **chmod**, then run the program to see the image and text on the window

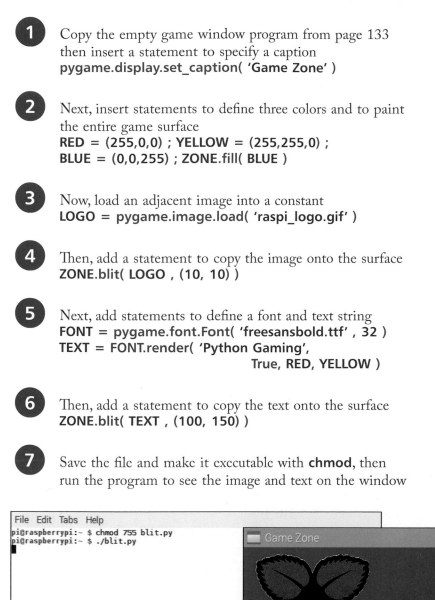

```
File Edit Tabs Help
pi@raspberrypi:~ $ chmod 755 blit.py
pi@raspberrypi:~ $ ./blit.py
```

Game Zone

Python Gaming

Playing sounds

The game loop that recognizes the **QUIT** event that occurs when the user ends the game can also recognize mouse clicks when the **MOUSEBUTTONDOWN** and **MOUSEBUTTONUP** events occur. Calling the **pygame.mouse.get_pos()** method in response to these events identifies the cursor position on the game surface. Testing the current position of the cursor enables the game surface to have defined "hot spots". So when the user clicks and is on a hot spot, the program can respond. For example, the response might be to play an audio file while the user continues to hold down the mouse button over a hot spot.

An audio file is made available to a game program by specifying it as the argument to the **pygame.mixer.Sound()** constructor to create a "Sound" object. This has **play()** and **stop()** methods that can be called in response to game events, such as a mouse click.

sound.py

freesansbold.ttf

1 Copy the empty game window program from page 133 then insert a statement to specify a caption
pygame.display.set_caption('Game Zone')

2 Next, insert statements to define three colors and to paint the entire game surface
RED = (255,0,0) ; YELLOW = (255,255,0) ;
BLUE = (0,0,255) ; ZONE.fill(BLUE)

3 Now, create three constants for a font, hot spot "button", and a fixed offset of 20 pixels for hot spot positioning
FONT = pygame.font.Font('freesansbold.ttf' , 32)
PLAY_BTN = FONT.render('Play', True, RED, YELLOW)
OFFSET = 20

4 Copy the hot spot "button" onto the game surface, with its top left corner 20 pixels from the top and left edges
ZONE.blit(PLAY_BTN, (OFFSET, OFFSET))

5 Then, add statements to calculate the position of each edge of the hot spot rectangle and store them in constants
PLAY_RECT = PLAY_BTN.get_rect()
BTN_L = PLAY_RECT.left + OFFSET
BTN_R = PLAY_RECT.right + OFFSET
BTN_TOP = PLAY_RECT.top + OFFSET
BTN_BTM = PLAY_RECT.bottom + OFFSET

The rectangle **get_rect()** method returns a rectangle that is the size of its surface dimensions.

6 Next, create a constant object specifying an audio file
SOUND = pygame.mixer.Sound('drum_loop.wav')

drum_loop.wav

7 Inside the game loop, insert a statement to respond by
playing the audio file if the user holds down the mouse
button when the cursor is over the hot spot
if event.type == MOUSEBUTTONDOWN :
        mouseX, mouseY = pygame.mouse.get_pos()
        **if mouseX >= BTN_L and mouseX <= BTN_R **
        **and mouseY >= BTN_TOP **
        and mouseY <= BTN_BTM : SOUND.play()

8 Inside the game loop, insert a statement to stop playing
the audio file when the user releases the mouse button
if event.type == MOUSEBUTTONUP :
        mouseX, mouseY = pygame.mouse.get_pos()
        **if mouseX >= BTN_L and mouseX <= BTN_R **
        **and mouseY >= BTN_TOP **
        and mouseY <= BTN_BTM : SOUND.stop()

9 Save the file and make it executable with **chmod**, then run
the program and click 'n' hold the mouse button on the
hot spot to hear the sound play

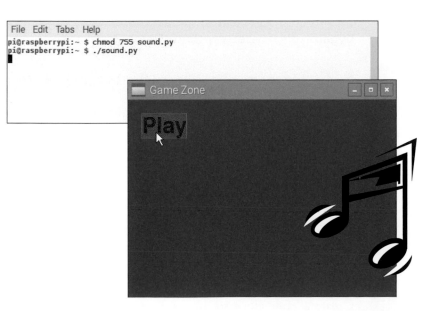

Having established the
cursor position when the
user clicks the mouse,
the program tests if it's
between the left and
right edges of the hot
spot and also between
its top and bottom
edges.

Moving images

A game program can interact with the user by recognizing when keyboard keys are pressed. The **pygame.key.get_pressed()** method returns a sequence of Boolean values representing the state of all keys, so can be tested to establish if a particular key is pressed. Usefully, each keyboard key has an associated constant value for this purpose. For example, the left arrow key has **K_LEFT**.

Typically, an image might move on the game surface in response to the user pressing an arrow key. This is simply achieved by modifying its coordinates in the game loop – so on each iteration the image gets repositioned while the key remains pressed.

A modifier variable can be assigned a value to determine the amount by which to reposition the image on each iteration and be assigned a zero value to halt movement.

The speed at which the image moves is the "frame rate" at which the game loop repaints the game surface. This will often be faster than is desirable, but can be limited by introducing a "clock" object into the game. This provides a **tick()** method to specify, within its parentheses, how many times per second the game loop should repaint the game surface – thereby controlling the speed of movement. A frame rate of 30 per second is generally preferable.

keymove.py

GIF

saucer.gif

1 Copy the empty game window program from page 133 then insert a statement to specify a caption
pygame.display.set_caption('Game Zone')

2 Next, insert statements to define color and image objects
BLUE = (0,0,255)
SAUCER = pygame.image.load('saucer.gif')

3 Now, initialize coordinate variables to set the starting position for the image rectangle's top left corner
x = y = 10

4 Then, initialize modifier variables at zero so there will be no movement steps when the game starts
mod_x = mod_y = 0

5 Create a game clock so the frame rate can be restricted
clock = pygame.time.Clock()

...cont'd

6 Inside the game loop, at the very beginning, insert statements to apply the movement steps, paint the game surface, and display the image at the calculated position

```
x += mod_x
y += mod_y
ZONE.fill( BLUE )
ZONE.blit( SAUCER, ( x,y ) )
```

When no arrow key is pressed, the program is still repainting the surface 30 times per second but no movement is visible.

7 Inside the game loop, in the **for event** section, insert statements to change the movement step values from zero in the event that any arrow key is pressed

```
keys = pygame.key.get_pressed()
if keys[ pygame.K_RIGHT ] : mod_x = 3
elif keys[ pygame.K_LEFT ] : mod_x = -3
elif keys[ pygame.K_UP ] : mod_y = -3
elif keys[ pygame.K_DOWN ] : mod_y = 3
else : mod_x = mod_y = 0
```

8 Finally, at the very end of the game loop, insert a statement to set the frame rate to 30 frames per second

```
clock.tick( 30 )
```

9 Save the file and make it executable with **chmod**, then run the program and use the arrow keys to move the image

```
File  Edit  Tabs  Help
pi@raspberrypi:~ $ chmod 755 keymove.py
pi@raspberrypi:~ $ ./keymove.py
```

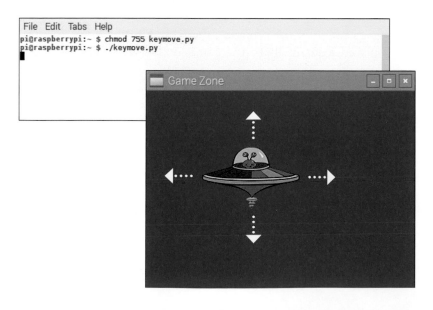

Press a horizontal and vertical arrow key together to set both modifier variables with a value – to see the image move diagonally.

Animating sprites

As the game loop constantly repaints the game surface while a program is running, animation is simply achieved by having slightly different images painted on each iteration. Rather than create separate image objects for each "frame" it is more efficient to produce a single sheet of slightly different images in sections then have the program display a section for each frame. Each item in the animation is known as a "sprite" and an entire sprite sheet looks like this:

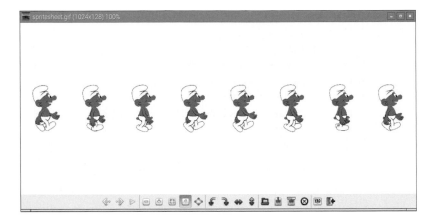

Note that this sprite sheet is 1024 x 128 pixels and contains 8 images – each one 128 x 128 pixels in size.

An image object for the entire sheet is created as usual with the **pygame.image.load()** method, and transparency can be preserved by appending **convert_alpha()** to the loading statement. Each section can then be assigned to a list by stating its coordinates within the sheet to the image object's **subsurface()** method. During execution of the program, the section in the list elements are displayed sequentially by the game loop to create animation:

animation.py

 1 Copy the empty game window program from page 133 then insert a statement to specify a caption
pygame.display.set_caption('Game Zone')

2 Next, insert statements to define color and clock objects
RED = (255, 0, 0)
clock = pygame.time.Clock()

3 Now, create an animation counter variable and an empty list to hold the spritesheet sections
counter = 0
sprites = []

4 Create the image object and get its width
```
sheet = \
pygame.image.load( 'spritesheet.gif' ).convert_alpha()
width = sheet.get_width()
```

spritesheet.gif

5 Next, assign the sprite sheet sections to the list
```
for i in range( int( width / 128 ) ) :
        sprites.append( \
                sheet.subsurface( i*128, 0, 128, 128 ) )
```

6 Now, in the game loop, add statements to paint the game surface on each iteration
```
ZONE.fill( RED )
ZONE.blit( sprites[ counter ] , (10,10) )
```

7 Then, increment the counter until it has displayed the final section then have it start over
```
counter = (counter + 1 ) % 8
```

8 Finally, at the very end of the game loop insert a statement to set the frame rate to 16 frames per second
```
clock.tick( 16 )
```

9 Save the file and make it executable with **chmod**, then run the program to see the animation

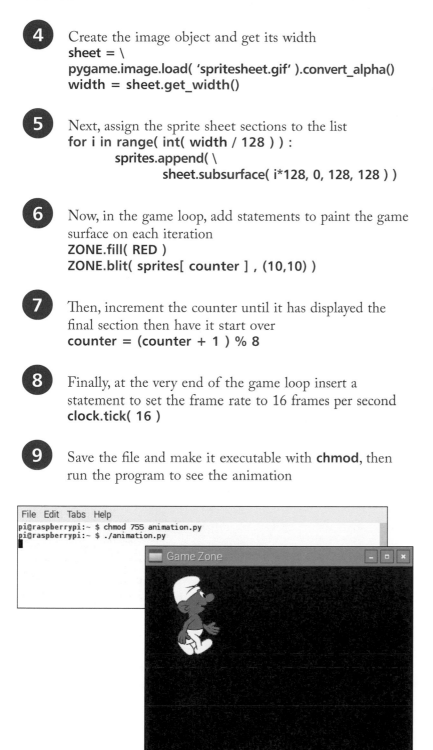

```
File  Edit  Tabs  Help
pi@raspberrypi:~ $ chmod 755 animation.py
pi@raspberrypi:~ $ ./animation.py
```

Game Zone

The clock sets a lower frame rate than the previous example because this animation needs to run slower.

143

Detecting collisions

Games will typically have multiple sprites and require detection of their collisions in order to implement a response. For example, to display an explosion animation sequence when a cannonball sprite collides with a ship sprite. In program terms, objects painted on the game surface are contained in bounding rectangles (even circles) which have a **colliderect()** method that can be used to detect when they overlap. Specifying the name of another object within this method's parentheses will return **True** when they are overlapping. This can then be used to provide a response such as changing the sprites' direction of movement:

collision.py

 Copy the empty game window program from page 133 then insert a statement to specify a caption
pygame.display.set_caption('Game Zone')

2 Next, insert statements to define three colors and a clock object to control animation speed
RED = (255, 0, 0) ; GREEN = (0, 255, 0) ;
BLUE = (0, 0, 255) ; clock= pygame.time.Clock()

3 Now, initialize coordinate and step variables for a red "ball" (circle) that will "bounce" around the game surface
rx = ry = 20
rdx = rdy = 1

The variable names in this example are undescriptive due to page space restrictions, but for **rx** think red_ball_position_x and for **rdx** think red_ball_direction_x.

4 Initialize coordinate and step variables for a green "ball" (circle) that will also "bounce" around the game surface
gx = gy = 280
gdx = gdy = -1

5 Then, initialize two variables with bounce points at which the balls will change direction to stay in the game surface
w = ZONE.get_rect().width - 10
h = ZONE.get_rect().height - 10

6 Now, in the game loop add statements to bounce the red ball by reversing its step value at the bounce points
if ((rx+rdx) > w) or ((rx+rdx) < 10) : rdx = -rdx
if ((ry+rdy) > h) or ((ry+rdy) < 10) : rdy = -rdy
rx += rdx ; ry += rdy

7 Similarly, add statements to bounce the green ball
```
if ( (gx+gdx) > w ) or ( (gx+gdx) < 10 ) : gdx = -gdx
if ( (gy+gdy) > h ) or ( (gy+gdy) < 10 ) : gdy = -gdy
gx += gdx
gy += gdy
```

8 Next, in the game loop, add statements to paint the game surface and the balls on each iteration
```
ZONE.fill( BLUE )
RED_BALL = \
        pygame.draw.circle( ZONE, RED, (rx,ry), 20 )
GREEN_BALL = \
        pygame.draw.circle( ZONE, GREEN, (gx,gy), 20 )
```

8 Then, in the game loop add statements to change direction when the balls collide
```
if RED_BALL.colliderect( GREEN_BALL ) :
        gdx = 1
        rdx = -1
```

9 Finally, at the very end of the game loop insert a statement to set the frame rate to 100 frames per second
```
clock.tick( 100 )
```

10 Save the file and make it executable with **chmod**, then run the program to see the collisions get detected

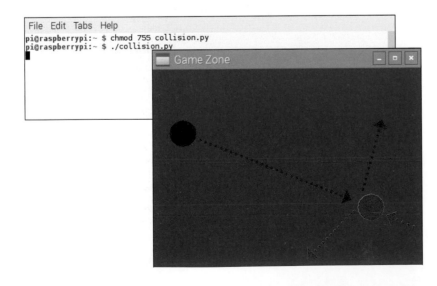

You can share games you create with other Raspberry Pi users via the "Pi Store" app. Install this with the commands **sudo apt-get update sudo apt-get install pistore**.

Summary

- The **pygame** module can be imported into a Python program to provide attributes and methods with which to create games.
- Every **pygame** program must begin by calling **pygame.init()** and create a game surface with **pygame.display.set_mode()**.
- The **while True** game loop is used to capture events that occur, such as **QUIT**, and can repaint the surface for animation.
- A game caption is set with **pygame.display.set_caption()**.
- Shapes are painted onto the surface using the **rect()**, **circle()**, **polygon()**, **line()** and **ellipse()** methods of **pygame.draw**.
- An image object is created by **pygame.image.load()** and can then be copied onto the surface using its **blit()** method.
- A font object is created by **pygame.font.Font()** which can then create a text block using its **render()** method.
- The **MOUSEBUTTONUP** and **MOUSEBUTTONDOWN** events get a current cursor position with **pygame.mouse.get_pos()**.
- A sound object is created by **pygame.mixer.Sound()** and has **play()** and **stop()** methods to control audio playback.
- The **pygame.key.get_pressed()** method returns a sequence of Boolean values representing the state of all keys.
- Each keyboard key has an associated constant such as **K_LEFT**.
- A game clock is created by **pygame.time.Clock()** and has a **tick()** method to specify a frame-per-second rate.
- Sprites are items in an animation that can be loaded into a list from a sheet using the image object's **subsurface()** method.
- Objects are painted on the surface in bounding rectangles and their **colliderect()** method can detect when they overlap.

8

Developing windowed apps

This chapter demonstrates how to create graphical windowed apps with the Python Tkinter module.

Introducing Tkinter

Raspberry Pi ships with the standard Python module that you can use to create graphical applications. It is called "Tkinter" – a **T**ool**k**it to **inter**face with the system GUI.

The **Tkinter** module can be imported into a program to provide attributes and methods for windowed apps. This module automatically imports another module of **Tkconstants** so all its constants are also then readily available to the program.

Every Tkinter program must begin by calling the **Tk()** constructor to create a window object. There must also be a call to the window object's **mainloop()** method to capture events, such as when the user closes the window to quit the program. This loop should appear at the end of the program, as it also handles window updates that may be implemented during execution.

tk_window.py

There can be only one call to the **Tk()** constructor and it must be at the start of the program code.

 1 Launch a plain text editor and begin a new Python program by locating the interpreter
#! /usr/bin/env python

2 Next, import the Tkinter module to make its attributes and methods available
from Tkinter import *

3 Now, create a window object by calling the constructor
window = Tk()

4 Then, add the loop to capture this window's events
window.mainloop()

5 Save the file then right-click on its graphical file icon and choose the **Properties** item on the context menu – to launch the File Properties dialog

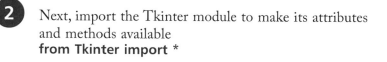

6 In the File Properties dialog, select the **Permissions** tab then check the **Execute** option for "Anyone"

You can of course use **chmod** to make the file executable as usual, but this example demonstrates the GUI alternative technique.

7 Click **OK** to close the File Properties dialog and to implement the change making the file executable

8 Now, double-click the file icon and push the **Execute** button on the dialog that appears to run the program – the dialog closes and an empty app window appears

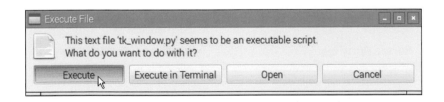

Double-click the file icon then push the **Open** button to open the source code in the Leafpad text editor.

9 Try out the minimize and maximize buttons, to see the window already has those features, then click the **X** button to quit the program and see the window close

Adding widgets

With **Tkinter** all the graphical controls that can be included in the application window, such as buttons or checkboxes, are referred to as "widgets". Each widget is listed in the table below, together with a brief description:

The **tkMessageBox** widget is defined in its own module which must be imported separately. It is demonstrated on page 154 later in this chapter.

Widget:	Description:
Button	Push button the user can click
Canvas	Area on which to draw shapes
Checkbutton	Checkbox options for multiple selections
Entry	Single-line text field for user input
Frame	Container for other widgets
Label	Single-line text display
Listbox	List of options for selection
Menubutton	Button to reveal a drop-down list
Menu	List revealed by Menubutton
Message	Multi-line text display
Radiobutton	Radio options allowing a single selection
Scale	Slider that the user can drag
Scrollbar	Scrollbar that the user can drag
Text	Multi-line text field for user input
Toplevel	Separate window container
Spinbox	Selector from a fixed range for input
PanedWindow	Container for widgets inside panes
LabelFrame	Spacer for complex layouts
tkMessageBox	Pop up message box

The default window's title of "tk", as in the previous example, can be changed by specifying a string within the parentheses of the window object's **title()** method.

A label object is created by specifying the window name and text=string as arguments to a **Label()** constructor. The label can then be added to the window using its **pack()** method. Optionally, this can specify padding with padx=n and pady=n values.

1 Launch a plain text editor and begin a new Python program by locating the interpreter
#! /usr/bin/env python

tk_label.py

2 Next, import the Tkinter module to make its attributes and methods available
from Tkinter import *

3 Now, create a window object by calling the constructor
window = Tk()

4 Specify a title to replace the default window title
window.title('Label Example')

5 Then, create a label object
label = Label(window , text = 'Hello World!')

6 Add the label to the window with both horizontal and vertical padding
label.pack(padx = 100 , pady = 50)

151

Padding values can also be specified as a tuple to apply the padding unequally. For example with **padx = (20, 80)**.

7 Finally, add the loop to capture this window's events
window.mainloop()

8 Save the file and make it executable, then run the program to see an application window appear bearing the specified title and the padded label

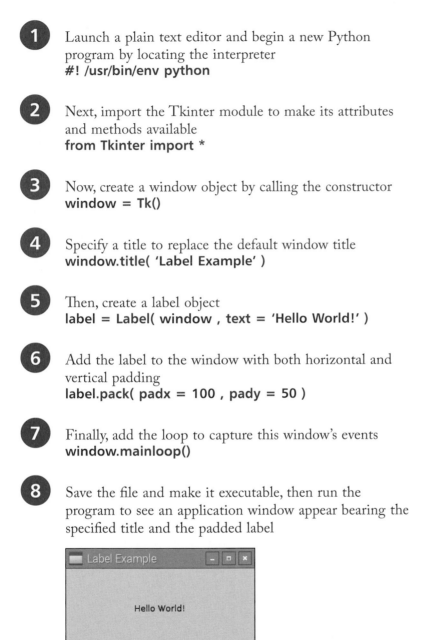

Notice that the padding has been applied all around the text, with 100 pixels each side horizontally and 50 pixels each side vertically, so the text is centered. The overall window size automatically assumes the proportions of its contents, so in this case the window fits snugly around the label.

Responding to buttons

A Button widget provides a graphical button in an application window that may contain either text or an image to convey the button's purpose. A button object is created by specifying the window name and options as arguments to a **Button()** constructor. Each option is specified as an option=value pair. The **command** option must always specify the name of a function or method to call when the user clicks that button. The most popular options are listed below, together with a brief description:

You can also call a button's **invoke()** method to, in turn, call the function nominated to its **command** option.

152

Option:	Description:
activebackground	Background color when the cursor over
activeforeground	Foreground color when the cursor over
bd	Border width in pixels (default is 2)
bg	Background color
command	Function to call when clicked
fg	Foreground color
font	Font for button label
height	Button height in text lines, or pixels
highlightcolor	Border color when in focus
image	Image to be displayed instead of text
justify	Multiple text lines LEFT,CENTER,RIGHT
padx	Horizontal padding
pady	Vertical padding
relief	Border SUNKEN,RIDGE,RAISED,GROOVE
state	Enabled status of NORMAL or DISABLED
underline	Index number of character underline
width	Button width in letters, or pixels
wraplength	Length at which to wrap text

The values assigned to other options determine the widget's appearance. These can be altered by specifying a new option=value pair as an argument to the widget's **configure()** method. Additionally, a current option value can be retrieved by specifying its name as a string argument to the widget's **cget()** method.

1 Begin a new Python program by locating the interpreter
and importing the Tkinter module attributes and methods

```
#! /usr/bin/env python
from Tkinter import *
```

tk_button.py

2 Next, create a window object and specify a title

```
window = Tk()
window.title( 'Button Example' )
```

3 Now, create a button to exit the program when clicked

```
btn_end = \
Button( window , text = 'Close' , command=exit )
```

4 Add a function to toggle the window color when clicked

```
def tog() :
        if window.cget( 'bg' ) == 'yellow' :
                window.configure( bg = 'gray' )
        else :
                window.configure( bg = 'yellow' )
```

Only the function
name is specified to
the **command** option.
Do not add trailing
parentheses in the
assignment.

5 Then, create a button to call the function when clicked

```
btn_tog = \
Button( window , text = 'Switch' , command=tog )
```

6 Now, add the buttons to the window with padding

```
btn_end.pack( padx = 100 , pady = 20 )
btn_tog.pack( padx = 100 , pady = 20 )
```

7 Finally, add the loop to capture this window's events

```
window.mainloop()
```

8 Save the file and make it executable, then run the program
and click the buttons to call their command functions

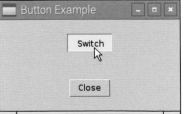
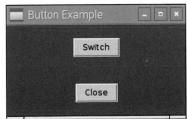

The 'gray' color is the
original default color of
the window.

153

Producing messages

A program can display messages to the user by calling methods provided in the "tkMessageBox" module. This must be imported separately and, as it has a lengthy name, an **import as** statement can usefully provide a short alias for use when calling its methods.

A message box is created by supplying a box title and the message to be displayed as the two arguments to one of these methods:

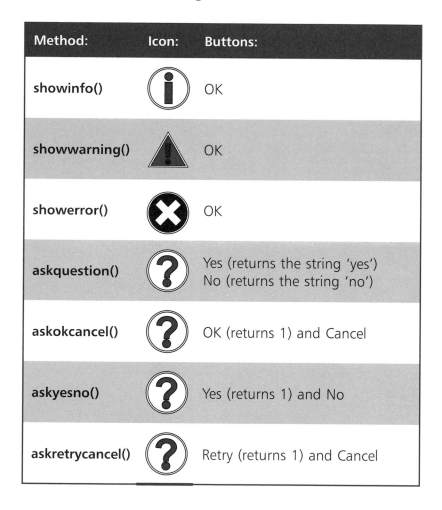

Method:	Icon:	Buttons:
showinfo()		OK
showwarning()		OK
showerror()		OK
askquestion()		Yes (returns the string 'yes') No (returns the string 'no')
askokcancel()		OK (returns 1) and Cancel
askyesno()		Yes (returns 1) and No
askretrycancel()		Retry (returns 1) and Cancel

Those methods that produce a message box containing a single OK button return no value when the button gets clicked by the user. Those that do return a value can be used to perform conditional branching by testing that value.

Only the **askquestion()** method returns two values – the **askyesno()** No button and both Cancel buttons return nothing.

Hot tip

1 Begin a new Python program by locating the interpreter and importing the Tkinter module attributes and methods
```
#! /usr/bin/env python
from Tkinter import *
```

tk_message.py

2 Also import the tkMessageBox module as a short alias
```
import tkMessageBox as box
```

3 Next, create a window object and specify a title
```
window = Tk()
window.title( 'Message Box Example' )
```

4 Add a function to display various message boxes
```
def dialog() :
        var = box.askyesno( 'Message Box' , 'Proceed?' )
        if var == 1 :
                box.showinfo( 'Box', 'Proceeding...' )
        else :
                box.showwarning( 'Box', 'Cancelling...'
)
```

Options can be added as a third argument to these method calls. For example, add **type = 'abortretryignore'** to get three buttons.

5 Then, create a button to call the function when clicked
```
btn = \
Button( window , text = 'Click' , command=dialog )
```

6 Now, add the button to the window with padding
```
btn.pack( padx = 100 , pady = 50 )
```

7 Finally, add the loop to capture this window's events
```
window.mainloop()
```

8 Save the file and make it executable, then run the program and click the button to see the message boxes

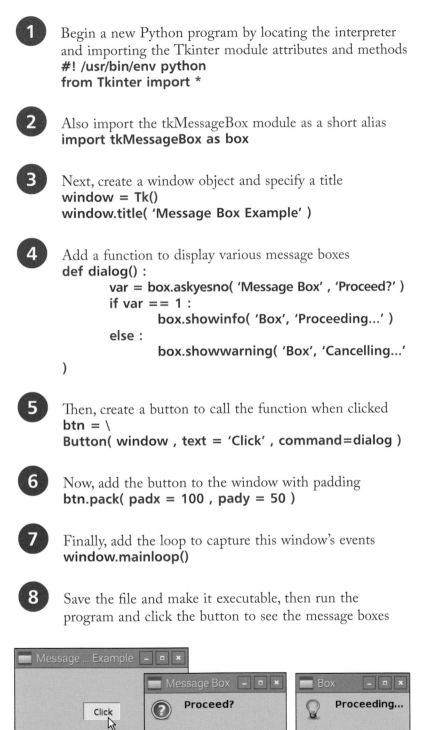

Gathering entries

An Entry widget provides a single-line input field in an application where the program can gather entries from the user. An entry object is created by specifying the name of its parent container, such as a window or frame name, and options as arguments to an **Entry()** constructor. Each option is specified as an option=value pair. Popular options are listed below, together with a brief description:

Option:	Description:
bd	Border width in pixels (default is 2)
bg	Background color
fg	Foreground color to render the text
font	Font for the text
highlightcolor	Border color when in focus
selectbackground	Background color of selected text
selectforeground	Foreground color of selected text
show	Hide password characters, show='*'
state	Enabled status of NORMAL or DISABLED
width	Entry width in letters

Use the Text widget instead of an Entry widget if you want to allow the user to enter multiple lines of text.

Multiple widgets can be grouped in frames for better positioning. A frame object is created by specifying the name of the window to a **Frame()** constructor. The frame's name can then be specified as the first argument to the widget constructors, to identify it as that widget's container.

When actually adding widgets to the frame, you can specify which side to pack them to in the frame with **TOP**, **BOTTOM**, **LEFT**, or **RIGHT** constants. For example, **entry.pack(side=LEFT)**.

Typically, an entry widget will appear alongside a label describing the type of input expected there from the user, or alongside a button widget that the user can click to perform some action on the data they have entered, so positioning in a frame is ideal.

Data currently entered into an entry widget can be retrieved by the program using that widget's **get()** method.

...cont'd

1 Begin a new Python program by locating the interpreter and importing the Tkinter module attributes and methods
```
#! /usr/bin/env python
from Tkinter import *
```

tk_entry.py

2 Also import the tkMessageBox module as a short alias
```
import tkMessageBox as box
```

3 Next, create a window object and specify a title
```
window = Tk()
window.title( 'Entry Example' )
```

4 Now, add a function to display data currently entered
```
def dialog() :
        box.showinfo('Greetings','Welcome '+entry.get())
```

5 Then, create a frame to contain widgets
```
frame = Frame( window )
```

6 Next, create an entry field for user input and a button to call the function when clicked
```
entry = Entry( frame )
btn = \
Button( frame, text = 'Enter Name', command=dialog )
```

Use a Label widget instead of an Entry widget if you want to display text that the user cannot edit.

7 Add the button and entry to the frame at set sides
```
btn.pack( side = RIGHT , padx = 5 )
entry.pack( side = LEFT )
frame.pack( padx = 20, pady = 20 )
```

8 Finally, add the loop to capture this window's events
```
window.mainloop()
```

9 Save the file and make it executable, then run the program and enter your name – click the button to see input used

157

Listing options

A Listbox widget provides a list of items in an application from which the user can make a selection. A listbox object is created by specifying the name of its parent container, such as a window or frame name, and options as arguments to a **Listbox()** constructor. Popular options are listed below, together with a brief description:

Option:	Description:
bd	Border width in pixels (default is 2)
bg	Background color
fg	Foreground color to render the text
font	Font for the text
height	Number of lines in list (default is 10)
selectbackground	Background color of selected text
selectmode	SINGLE (default) or MULTIPLE selections
width	Listbox width in letters (default is 20)
yscrollcommand	Attach to a vertical scrollbar

With Tkinter, a scrollbar is a separate widget that can be attached to Listbox, Text, Canvas and Entry widgets.

Items are added to the listbox by specifying a list index number and the item string as arguments to its **insert()** method.

You can retrieve any item from a listbox by specifying its index number within the parentheses of its **get()** method. Usefully, a listbox also has a **curselection()** method that returns the index number of the currently selected item, so this can be supplied as the argument to its **get()** method to retrieve the current selection:

tk_listbox.py

 Begin a new Python program by locating the interpreter and importing the Tkinter module attributes and methods
```
#! /usr/bin/env python
from Tkinter import *
```

 Also import the tkMessageBox module as a short alias
```
import tkMessageBox as box
```

Next, create a window object and specify a title
```
window = Tk()
window.title( 'Listbox Example' )
```

4 Now, add a function to display a listbox selection
```
def dialog() :
        box.showinfo( 'Selection', 'Your Choice: ' + \
        listbox.get( listbox.curselection() ) )
```

5 Then, create a frame to contain widgets
```
frame = Frame( window )
```

6 Next, create a listbox widget offering three items
```
listbox = Listbox( frame )
listbox.insert( 1, 'HTML5 in easy steps' )
listbox.insert( 2, 'CSS3 in easy steps' )
listbox.insert( 3, 'JavaScript in easy steps' )
```

7 Now, create a button to call the function when clicked
```
btn = Button( frame,text = 'Choose',command=dialog )
```

8 Add the button and listbox to the frame at set sides
```
btn.pack( side = RIGHT , padx = 5 )
listbox.pack( side = LEFT )
frame.pack( padx = 20, pady = 20 )
```

9 Finally, add the loop to capture this window's events
```
window.mainloop()
```

10 Save the file and make it executable, then run the program and choose a list item to see your selection confirmed

If the **selectmode** is set to MULTIPLE, the **curselection()** method returns a tuple of the selected index numbers.

You cannot use a regular variable to store values assigned from a radio button selection – it must be an object.

Polling radio buttons

A Radiobutton widget provides a single item in an application that the user may select. Where a number of radio buttons are grouped together, the user may only select any one item in the group. With Tkinter, radio button objects are grouped together when they nominate the same control variable object to assign a value to upon selection. An empty string variable object can be created for this purpose using the **StringVar()** constructor or an empty integer variable object using the **IntVar()** constructor.

A radio button object is created by specifying four arguments to a **Radiobutton()** constructor:

- Name of the parent container, such as the frame name
- Text for a display label, specified as a **text=**text pair
- Control variable object, specified as a **variable=**variable pair
- Value to be assigned, specified as a **value=**value pair

A string value assigned by selecting a radio button can be retrieved from a string variable object by its **get()** method.

tk_radio.py

 Begin a new Python program by locating the interpreter and importing the Tkinter module attributes and methods
#! /usr/bin/env python
from Tkinter import *

 Also import the tkMessageBox module as a short alias
import tkMessageBox as box

 Next, create a window object and specify a title
window = Tk()
window.title('Radio Button Example')

 Now, add a function to display a radio button selection
def dialog() :
    **box.showinfo('Selection', **
    'Your Choice: \n' + book.get())

5 Then, create a frame to contain widgets
frame = Frame(window)

Next, construct a string variable object to store a selection
book = StringVar()

 Now, create three radio button widgets whose value will be assigned to the string variable upon selection
radio_1 = Radiobutton(frame, text = 'HTML5',
** variable = book, value = 'HTML5 in easy steps')**
radio_2 = Radiobutton(frame, text = 'CSS3',
** variable = book, value = 'CSS3 in easy steps')**
radio_3 = Radiobutton(frame, text = 'JS',
** variable = book, value = 'JavaScript in easy steps')**

A Radiobutton object has a **select()** method that can be used to specify a default selection. For example, **radio_1.select()**.

7 Create a button to call the function when clicked
btn = Button(frame, text='Choose', command=dialog)

8 Then, add the push button and radio buttons to the frame
btn.pack(side = RIGHT , padx = 5)
radio_1.pack(side = LEFT)
radio_2.pack(side = LEFT)
radio_3.pack(side = LEFT)
frame.pack(padx = 20, pady = 20)

9 Finally, add the loop to capture this window's events
window.mainloop()

 Save the file and make it executable, then run the program and choose a radio button to see your selection confirmed

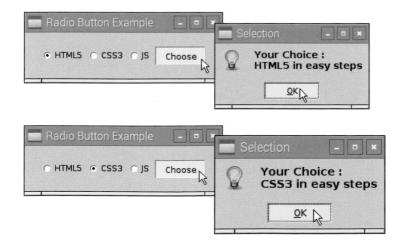

A Radiobutton object has a **deselect()** method that can be used to cancel a selection programmatically.

Checking boxes

A Checkbutton widget provides a single item in an application that the user may select. Where a number of check buttons appear together, the user may select one or more items. Check button objects nominate an individual control variable object to assign a value to, whether checked or unchecked. An empty string variable object can be created for this using the **StringVar()** constructor or an empty integer variable object using the **IntVar()** constructor.

A check button object is created by specifying five arguments to a **Checkbutton()** constructor:

- Name of the parent container, such as the frame name
- Text for a display label, as a **text=**text pair
- Control variable object, as a **variable=**variable pair
- Value to assign if checked, as an **onvalue=**value pair
- Value to assign if unchecked, as an **offvalue=**value pair

An integer value assigned by a check button can be retrieved from a integer variable object by its **get()** method.

tk_check.py

1 Begin a new Python program by locating the interpreter and importing the Tkinter module attributes and methods
```
#! /usr/bin/env python
from Tkinter import *
```

2 Also import the tkMessageBox module as a short alias
```
import tkMessageBox as box
```

3 Next, create a window object and specify a title
```
window = Tk()
window.title( 'Check Button Example' )
```

4 Now, add a function to display a check button selection
```
def dialog() :
    str = 'Your Choice:'
    if var_1.get()==1 : str+='\nHTML5 in easy steps'
    if var_2.get()==1 : str+='\nCSS3 in easy steps'
    if var_3.get()==1 : str+='\nJavaScript in easy steps'
    box.showinfo( 'Selection' , str )
```

5 Then, create a frame to contain widgets
```
frame = Frame( window )
```

6 Construct three integer variable objects to store values
```
var_1 = IntVar() ; var_2 = IntVar() ; var_3 = IntVar()
```

7 Create three check button widgets whose values will be assigned to the integer variable, whether checked or not
```
book_1 = Checkbutton( frame, text = 'HTML5',
  variable = var_1, onvalue = 1, offvalue = 0 )
book_2 = Checkbutton( frame, text = 'CSS3',
  variable = var_2, onvalue = 1, offvalue = 0 )
book_3 = Checkbutton( frame, text = 'JS',
  variable = var_3, onvalue = 1, offvalue = 0 )
```

A Checkbutton object has **select()** and **deselect()** methods that can be used to turn the state on or off.
e.g. **check_1.select()**.

8 Now, create a button to call the function when clicked
```
btn = Button( frame, text='Choose', command=dialog )
```

9 Then, add the push button and check buttons to the frame
```
btn.pack( side = RIGHT , padx = 5 )
book_1.pack( side = LEFT )
book_2.pack( side = LEFT )
book_3.pack( side = LEFT )
frame.pack( padx = 20, pady = 20 )
```

10 Finally, add the loop to capture this window's events
```
window.mainloop()
```

11 Save the file and make it executable, then run the program and choose check buttons to see your selection confirmed

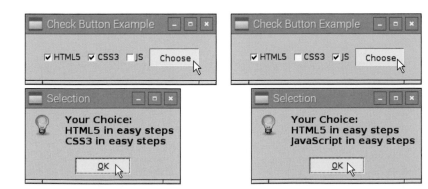

The state of any Checkbutton object can be reversed by calling its **toggle()** method.

The PhotoImage class also has a **zoom()** method that will double the image size with the same **x=2,y=2** values.

Displaying images

With Tkinter, images in GIF or PGM/PPM file formats can be displayed on Label, Button, Text and Canvas widgets using its **PhotoImage()** constructor to create image objects. This simply requires a single **file=** argument to specify the image file. Interestingly, it also has a **subsample()** method that can scale down a specified image by stating a sample value to **x=** and **y=** arguments. For example, values of **x=2, y=2** samples every second pixel – so the image object is half-size of the original.

Once an image object has been created it can be added to a Label or Button constructor statement by an **image=** option.

Text objects have an **image_create()** method with which to embed an image into the text field. This requires two arguments to specify location and **image=**. For example, **'1.0'** specifies the first line and first character.

Canvas objects have a **create_image()** method that requires two arguments to specify location and **image=**. Here, the location sets the X,Y coordinates on the canvas at which to paint the image.

tk_photo.py

raspi_logo.gif

1 Begin a new Python program by locating the interpreter and importing the Tkinter module attributes and methods
#! /usr/bin/env python
from Tkinter import *

2 Next, create a window object and specify a title
window = Tk()
window.title('Image Example')

3 Now, create an image object from a local image file
logo = PhotoImage(file = 'raspi_logo.gif')

4 Then, create a label object to display the image above a colored background
label = Label(window, image = logo, bg = 'yellow')

5 Create a half-size image object from the first image object
small_logo = PhotoImage.subsample(logo, x=2, y=2)

6 Now, create a button to display the small image
btn = Button(window, image = small_logo)

7 Create a text field and embed the small image, then insert some text after it
```
txt = Text( window, width = 30, height = 9 )
txt.image_create( '1.0', image = small_logo )
txt.insert( '1.1', 'Raspberry Pi' )
```

Notice that the Text method is **image_create()** but the Canvas method is **create_image()** – similar yet different.

8 Create a canvas and paint the small image above a colored background, then paint a diagonal line over the top of it
```
can = \
Canvas(window, width=120, height=120, bg='cyan')
can.create_image( ( 60, 60 ), image = small_logo )
can.create_line( 0,0,120,120, width=30, fill='yellow' )
```

9 Then, add the widgets to the window
```
label.pack( side = TOP )
btn.pack( side = LEFT, padx = 10 )
txt.pack( side = LEFT )
can.pack( side = LEFT, padx = 10 )
```

10 Finally, add the loop to capture this window's events
```
window.mainloop()
```

11 Save the file and make it executable, then run the program to see the widgets displaying an image

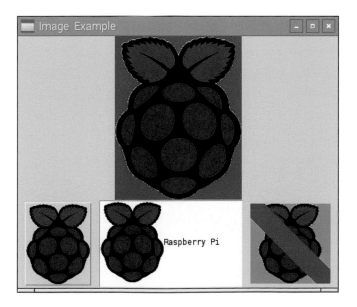

Text and Canvas widgets are both powerful and flexible – discover more online at **docs.python. org/2/library/ tkinter.html**

Adjusting attributes

Tkinter widgets often have many option attributes to specify their behavior and appearance but some attributes of the same name set different properties. In a Canvas widget, for example, the numeric values assigned to **width** and **height** attributes specify the widget's dimensions in pixels, whereas in a Label widget the values assigned to its **width** and **height** attributes specify the widget's dimensions as the number of characters and lines respectively.

The **text** value assigned to a Label widget will, by default, be positioned centrally within that label. Alternative positioning is possible by assigning constant compass points to an anchor attribute. For example, **anchor=N** positions text at top center.

Colors can be assigned to attributes as a hexadecimal string representation of their Red, Green, and Blue proportions. For example, '#000000' is black, '#FFFFFF' is white, '#FF0000 is red. Alternatively, you can use the standard color name strings **'black'**, **'white'**, **'red'**, **'green'**, **'blue'**, **'yellow'**, **'magenta'**, and **'cyan'**.

A widget can be surrounded by a simulated 3D effect specified to its **relief** attribute with constants of **FLAT**, **RAISED**, **SUNKEN**, **GROOVE**, or **RIDGE**. By default, a Button widget gets the **RAISED** effect, but an alternative can be specified to its **relief** attribute.

The appearance of the cursor can be changed when placed over a widget by specifying an alternative to its **cursor** attribute using one of the values listed below:

- 'arrow'
- 'circle'
- 'clock'
- 'cross'
- 'dotbox'
- 'exchange'
- 'fleur'

- 'heart'
- 'man'
- 'mouse'
- 'pirate'
- 'plus'
- 'shuttle'
- 'sizing'

- 'spider'
- 'spraycan'
- 'star'
- 'target'
- 'tcross'
- 'trek'
- 'watch'

tk_attrib.py

1 Begin a new Python program by locating the interpreter and importing the Tkinter module attributes and methods
#! /usr/bin/env python
from Tkinter import *

2 Next, create a window object and specify a title
```
window = Tk()
window.title( 'Attribute Example' )
```

3 Now, create five labels with lots of options
```
lbl_1 = Label( window,text='A',anchor=N,width=8, \
height=5, bg='black', fg='white', cursor='circle' )

lbl_2 = Label( window,text='B',anchor=S,width=8, \
height=5, bg='red', fg='white', cursor='pirate' )

lbl_3 = Label( window,text='C',anchor=E, width=8, \
height=5, bg='green', cursor='heart' )

lbl_4 = Label( window,text='D',anchor=NE,width=8, \
height=5, bg='yellow', fg='#FF0000', cursor='watch' )

lbl_5 = Label( window,text='E',anchor=SW,width=8, \
height=5, bg='cyan', cursor='spider' )
```

With a Label widget, the **width** attribute specifies a number of characters and the **height** attribute specifies number of lines.

4 Then, add the labels to the window
```
lbl_1.pack( side = LEFT, padx = 5, pady = 5 )
lbl_2.pack( side = LEFT , padx = 5)
lbl_3.pack( side = LEFT , padx = 5)
lbl_4.pack( side = LEFT , padx = 5)|
lbl_5.pack( side = LEFT , padx = 5)
```

5 Finally, add the loop to capture this window's events
window.mainloop()

6 Save the file and make it executable, then run the program and note the text – roll the mouse over to see the cursors

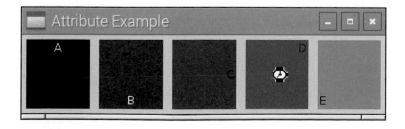

167

Managing layouts

The **pack()** method that has been used throughout this chapter to add widgets to applications is known as a "geometry manager". It places the widgets against a specified side of its parent container using the **TOP**, **BOTTOM**, **LEFT**, **RIGHT** constants and has a **fill** option to expand the widget in available space on a specified X or Y axis.

There is also a **grid()** geometry manager that places the widgets in cells, specified numerically to its **row** and **column** options, and a **place()** geometry manager that places the widgets at XY coordinates specified numerically to its **x** and **y** options.

tk_layout.py

The order of the pack statements ("packing order") matters. Change it around to see the difference. Most layout difficulties can be solved using the grid geometry manager instead of pack.

1 Begin a new Python program by locating the interpreter and importing the Tkinter module attributes and methods
#! /usr/bin/env python
from Tkinter import *

2 Next, create a window object and specify a title
window = Tk()
window.title('Layout Example')

3 Now, create four colored labels
lbl_red = Label(window,width=12,height=5,bg='red')
lbl_grn = Label(window,width=12,height=5,bg='green')
lbl_blu = Label(window,width=12,height=5,bg='blue')
lbl_yel =Label(window,width=12,height=5,bg='yellow')

4 Add the labels to the window, packed to different sides
lbl_red.pack(side = TOP)
lbl_grn.pack(side = BOTTOM)
lbl_blu.pack(side = LEFT)
lbl_yel.pack(side = RIGHT)

5 Finally, add the loop to capture this window's events
window.mainloop()

6 Save the file and make it executable, then run the program to see the pack layout

168

7 Edit the pack statement of the green label to make it fill the available space
**lbl_grn.pack(side=BOTTOM,\
                fill = X)**

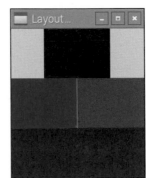

8 Save the file, then run the program to see the change

9 Next, edit the program to use the grid geometry manager, by changing all statements in step 4
**lbl_red.grid(row = 1, column = 1)
lbl_grn.grid(row = 2, column = 2)
lbl_blu.grid(row = 3, column = 3)
lbl_yel.grid(row = 1, column = 4)**

The grid geometry manager also has **rowspan** and **columnspan** options for you to create more elaborate grid layouts.

10 Save the file, then run the program to see the grid layout

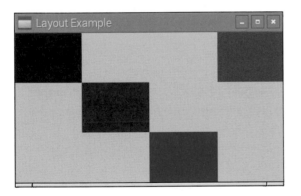

11 Now, edit the program to use the place geometry manager
**lbl_red.place(x=10, y=10)
lbl_grn.place(x=36, y=45)
lbl_blu.place(x=63, y=80)
lbl_yel.place(x=90, y=115)**

12 Save the file, then run the program to see the place layout

The three layout geometry managers (pack, grid, and place) should never be mixed in the same window!

Summary

- The **Tkinter** module can be imported into a Python program to provide attributes and methods for windowed applications.

- Every **Tkinter** program must begin by calling **Tk()** to create a window and call its **mainloop()** method to capture events.

- The window object's title is specified by its **title()** method.

- A label widget is created by specifying the name of its parent container and its text as arguments to the **Label()** constructor.

- A button widget is created by specifying the name of its parent container, its text, and the name of a function to call when the user pushes it, as arguments to the **Button()** constructor.

- The **tkMessageBox** module can be imported into a Python program to provide attributes and methods for message boxes.

- Message boxes that ask the user to make a choice return a value to the program for conditional branching.

- The **Frame()** constructor creates a container in which multiple widgets can be grouped for better positioning.

- The **Entry()** constructor creates a single line text field whose current contents can be retrieved by its **get()** method.

- Items are added to a **Listbox** object by its **insert()** method and retrieved by specifying their index number to its **get()** method.

- **Radiobutton** and **Checkbutton** objects store values in the **StringVar** or **IntVar** object nominated by their **variable** attribute.

- The **PhotoImage()** constructor creates an image object that has a **subsample()** method which can scale down the image.

- Images can be added to **Button** and **Label** objects, embedded in **Text** objects, and painted on **Canvas** objects.

- Widgets can be added to an application using the **pack()**, **grid()** or **place()** geometry managers.

9 Driving header pins

This chapter demonstrates how to control electrical input and output on the Raspberry Pi header from Python scripts.

Understanding pin numbering

The Raspberry Pi board incorporates a number of General Purpose Input Output (GPIO) header pins, whose behavior can be controlled by Python scripts. On the latest Raspberry Pi boards there are in total 40 header pins, arranged in two rows of 20 on a corner of the board:

The pins are made available to the system via connections directly on the BCM2837 chip. Although similar in appearance, the pins allow different electrical connections, so it is important to understand how each pin may be used.

Each pin can be referred to numerically. Pin number 1 ("P1") is at the extreme left on the bottom row in the photo above. This pin provides a 3.3 volt power supply limited to 50mA, and adjacent pin number 2 provides a 5 volt supply. You can work out each pin number from the pin map shown opposite. There are two numbering options provided on the map, as you can refer to the pin by either by its number on the board (1-40) or you can use the "BCM" name of the connection on the Broadcom system chip to which the pin is physically connected ("BCM 0"-"BCM 27").

Input and output can be controlled from a Python script for those pins that have BCM names on the pin map. For example, BCM3 (pin number 5) is programmable.

Connecting to the wrong pin may damage your Raspberry Pi, so extreme caution is advised using the GPIO header pins.

P1

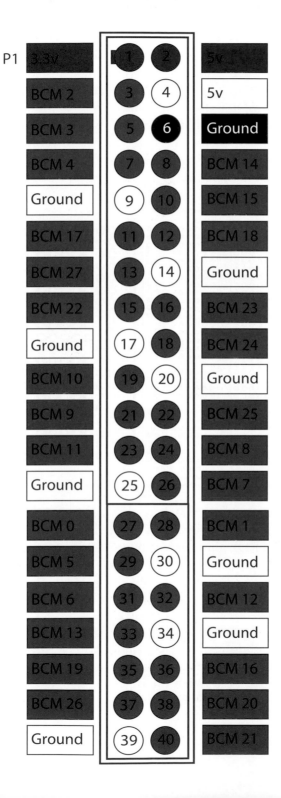

3.3v	1 · 2	5v
BCM 2	3 · 4	5v
BCM 3	5 · 6	Ground
BCM 4	7 · 8	BCM 14
Ground	9 · 10	BCM 15
BCM 17	11 · 12	BCM 18
BCM 27	13 · 14	Ground
BCM 22	15 · 16	BCM 23
Ground	17 · 18	BCM 24
BCM 10	19 · 20	Ground
BCM 9	21 · 22	BCM 25
BCM 11	23 · 24	BCM 8
Ground	25 · 26	BCM 7
BCM 0	27 · 28	BCM 1
BCM 5	29 · 30	Ground
BCM 6	31 · 32	BCM 12
BCM 13	33 · 34	Ground
BCM 19	35 · 36	BCM 16
BCM 26	37 · 38	BCM 20
Ground	39 · 40	BCM 21

Notice that one row of pins are even numbers and the other row are odd numbers.

Some earlier versions of the Raspberry Pi board have only 26 pins but later models extended the GPIO header to 40 pins. Happily, the pin numbering and BCM naming system is consistent throughout all Raspberry Pi boards – so any board can make identical use of the first 26 pins. The following examples in this chapter therefore use only these 26 pins to remain relevant to users of any Raspberry Pi board.

Lighting a lamp

When building projects for Raspberry Pi's GPIO header pins, it is convenient to use a "breadboard" to make connections. This allows you to easily plug in and remove components so is much quicker than soldering up the circuits. A breadboard has many holes that you can insert jumper leads/wires into that will be held in place by metal contacts inside. There are "power rail" columns of holes at the edges – their contacts are connected vertically. There are rows of holes either side of a central bridge – their contacts are connected horizontally:

Breadboards, LEDs, resistors and jumper leads can be purchased from most electronics retailers.

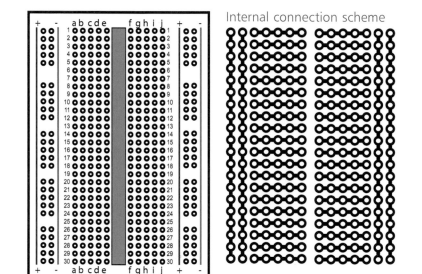

Internal connection scheme

- +

To get started with Raspberry Pi's GPIO pins you can make a simple circuit to light up a single "lamp" – a Light Emitting Diode (LED). These only allow electricity to pass in one direction so it is important to attach them to the breadboard the right way round. An LED has a long leg and a shorter leg. The long leg goes to the plus side, where the power is coming in.

If too much current is allowed to pass through the LED it will burn very brightly for a very short period of time then burn out. So a 270Ω (ohm) resistor must be added to the circuit to limit the current passing through the LED.

1 Turn off the power supply to your Raspberry Pi

2 Wire a 3.3v jumper lead power connection from GPIO pin 1 to the second row of the breadboard ▬

3 Next, insert an LED to connect the second and third rows, with the long leg inserted on the second row – where the power is coming in ●

4 Now, insert a resistor to connect the third row of the breadboard with the minus side of the power rail

5 Finally, wire a jumper lead connection from the minus side of the power rail to ground at GPIO pin 6 ▬ – your circuit should now look something like this:

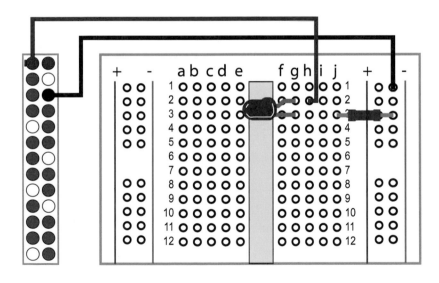

You can use rows and power rails on either side of the bridge, but not mixed. The illustrations use the rows and power rails on the right side only for clarity.

6 Turn on the power supply to the Raspberry Pi and you should instantly see the LED lamp illuminate

Directing output

Now that the simple circuit in the previous example established an illuminated LED, you can easily control its on/off status by switching the circuit's power supply to come from a programmable GPIO pin:

gpio_blink.py

1 Turn off the power supply to your Raspberry Pi

2 Disconnect the jumper lead power connection from GPIO pin 1, then connect it to pin 7 (BCM4)

3 Now, your circuit should look something like this:

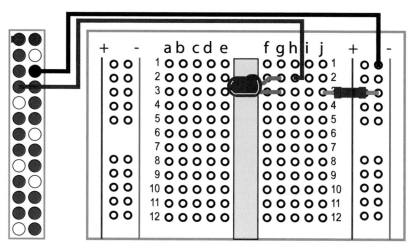

All Python scripts that address the GPIO pins must be run with superuser privileges – using a **sudo** command.

Raspberry Pi ships with a Python module called "RPi.GPIO" that contains attributes and methods to control input and output via the header pins. This can usefully be given a shorter alias with an **import as** statement.

The **RPi.GPIO** module provides a **setmode()** method to specify whether the script will address the header pins by their board number, specifying a **BOARD** constant argument, or by their chip number, specifying a **BCM** constant argument.

The **RPi.GPIO** module's **setup()** method requires two arguments to specify the number of a pin to be controlled and whether that pin will provide for output, specifying an **OUT** constant argument, or provide for input, specifying an **IN** constant argument. Channels set up for output can be reset for input by the **cleanup()** method.

The **RPi.GPIO** module's **output()** method requires two arguments to specify the number of a pin where output is to be directed and a Boolean value of **True** to provide power to that pin, or a Boolean value of **False** to deny power to that pin.

It is often useful with Python scripts for GPIO to import the **sleep()** method from the **time** module. This allows the script to make delays by the number of seconds specified as its argument.

The first 26 GPIO header pins include 17 pins that can be configured as inputs and outputs. By default, they are all configured as inputs except 8 & 10 (BCM 14 & BCM 15).

 Turn on your Raspberry Pi, then begin a new Python script by making attributes and methods available
import RPi.GPIO as GPIO
from time import sleep

 Add statements to use board pin numbers in the script and set up pin 7 to supply power to the LED as output
GPIO.setmode(GPIO.BOARD)
GPIO.setup(7 , GPIO.OUT)

This script could alternatively have used **GPIO.setmode(BCM)** ; **GPIO.setup(4,GPIO.OUT)** for BCM chip numbering.

 Now, add a loop to alternate the status of the pin three times, with a one second delay, then reset the channel
i = 1
while i < 4 :
      print('Cycle: ' + str(i))
      print('Set Output True - LED ON')
      GPIO.output(7 , True) ; sleep(1)
      print('Set Output False - LED OFF')
      GPIO.output(7 , False) ; sleep(1) ; i += 1
GPIO.cleanup()

 Save the file, then enter this command to run the script with superuser privileges and see the LED blink on and off
sudo python gpio_blink.py

```
pi@raspberrypi: ~
File  Edit  Tabs  Help
pi@raspberrypi:~ $ sudo python gpio_blink.py
Cycle: 1
        Set Output True - LED ON
        Set Output False - LED OFF
Cycle: 2
        Set Output True - LED ON
        Set Output False - LED OFF
Cycle: 3
        Set Output True - LED ON
        Set Output False - LED OFF
pi@raspberrypi:~ $
```

Adding more lamps

Now that the previous example controlled a single LED to make it blink on and off, you can add more LEDs to the circuit and have a Python script illuminate them, sequentially:

gpio_sequence.py

 Turn off the power supply to your Raspberry Pi

 Wire a jumper lead from pin 11 (BCM17) to the fourth row of the breadboard ▬

 Next, insert an LED to connect the fourth and fifth rows, with the long leg inserted on the fourth row ●

 Now, insert a resistor to connect the fifth row of the breadboard with the minus side of the power rail

 Wire a jumper lead power from pin 13 (BCM27) to the seventh row of the breadboard ▬

 Next, insert an LED to connect the seventh and eighth rows, with the long leg inserted on the seventh row ●

Finally, insert a resistor to connect the eighth row of the breadboard with the minus side of the power rail – your circuits should now look something like this:

Ensure the legs of the LEDs do not touch and create a short circuit.

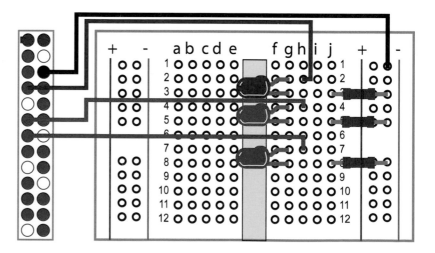

8 Turn on your Raspberry Pi then begin a new Python script by making attributes and methods available
import RPi.GPIO as GPIO
from time import sleep

9 Add statements to use board pin numbers in the script, and set up pins 7,11 and 13 to supply power as output
GPIO.setmode(GPIO.BOARD) ; GPIO.setup(7,GPIO.OUT)
GPIO.setup(11,GPIO.OUT) ; GPIO.setup(13,GPIO.OUT)

10 Now, add a loop to sequentially light each LED for one second on each of three iterations, then reset the channels

```
i = 1
while i < 4 :
        print( 'Cycle: ' + str( i ) )
        GPIO.output( 7 , True ) ;
        print( '\t7 Output True - RED ON' ) ; sleep( 1 )
        GPIO.output( 7 , False )
        GPIO.output( 11 , True ) ;
        print( '\t11 Output True - YELLOW ON' ) ; sleep(1)
        GPIO.output( 11 , False )
        GPIO.output( 713 , True ) ;
        print( '\t13 Output True - GREEN ON' ) ; sleep( 1 )
        GPIO.output( 13 , False )
        i += 1
GPIO.cleanup()
```

If you miss out the loop incrementer **i+=1** it will continue to run – press **Ctrl + C** to exit the script.

11 Save the file, then enter this command to run the script with superuser privileges and see the LED sequence
sudo python gpio_sequence.py

```
pi@raspberrypi: ~                                         _ □ ✕
File  Edit  Tabs  Help
pi@raspberrypi:~ $ sudo python gpio_sequence.py
Cycle: 1
        7 Output True - RED ON
        11 Output True - YELLOW ON
        13 Output True - GREEN ON
Cycle: 2
        7 Output True - RED ON
        11 Output True - YELLOW ON
        13 Output True - GREEN ON
Cycle: 3
        7 Output True - RED ON
        11 Output True - YELLOW ON
        13 Output True - GREEN ON
pi@raspberrypi:~ $ ▮
```

Recognizing input

The previous examples have controlled output to illuminate LEDs but you can also recognize input, such as a button depression, by creating a loop to "listen" for the button getting pushed.

The **RPi.GPIO** module's **setup()** method must specify the number of a pin to be used to listen for input and its **IN** argument.

The **RPi.GPIO** module's **input()** method simply requires the number of the pin that is listening as its sole argument. This method returns a Boolean value of **True** unless the button is pushed, then it returns a Boolean value of **False**. Testing for this condition change in a loop allows the script to respond when the button is pushed. First, you need to add a push button to the circuit:

gpio_button.py

 Turn off the power supply to your Raspberry Pi

 Next, wire a jumper lead from pin 5 (BCM3) to the ninth row of the breadboard ▬

 Now, insert a button to connect the ninth and tenth rows

4 Finally, wire a jumper lead to connect the tenth row of the breadboard with the minus side of the power rail ▬ – your circuits should now look something like this:

Choose jumper leads of different colors for each circuit for easy recognition.

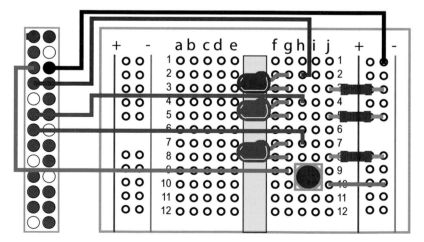

5 Turn on your Raspberry Pi then begin a new Python script by making attributes and methods available
import RPi.GPIO as GPIO
from time import sleep

Ensure the button is inserted the right way round – to connect rows, not columns!

6 Add statements to use board pin numbers in the script and set up pin 7 to supply power as output, then set up pin 5 to listen for input
GPIO.setmode(GPIO.BOARD)
GPIO.setup(7 , GPIO.OUT)
GPIO.setup(5 , GPIO.IN)

7 Now, add a loop to illuminate the LED for one second on each of three occasions when the button gets pushed, then reset the channel
```
i = 1
while i < 4 :
        if GPIO.input( 5 ) :
                GPIO.output( 7 , False )
        else :
                GPIO.output( 7 , True ) ;
                print( 'Button Pushed: ' + str( i ) )
                print( '\t7 Output True - RED ON' )
                sleep( 1 )
                i += 1
GPIO.cleanup()
```

Every iteration of the loop explicitly turns off the LED unless the button gets pushed when it turns the LED on for one second.

181

8 Save the file, then enter this command to run the script and push the button to see the LED light up in response
sudo python gpio_button.py

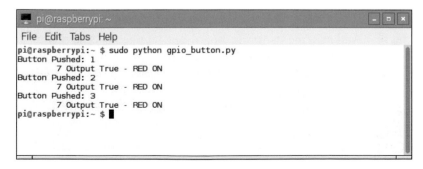

```
pi@raspberrypi: ~
File Edit Tabs Help
pi@raspberrypi:~ $ sudo python gpio_button.py
Button Pushed: 1
        7 Output True - RED ON
Button Pushed: 2
        7 Output True - RED ON
Button Pushed: 3
        7 Output True - RED ON
pi@raspberrypi:~ $
```

Adding more buttons

A second button circuit can now be added to the previous example so LED lamps can be turned on and off independently:

gpio_inputs.py

1 Turn off the power supply to your Raspberry Pi then wire a jumper lead from pin 3 (BCM2) to the eleventh row of the breadboard

2 Insert a button to connect the eleventh and twelfth rows then wire a jumper lead to connect the twelfth row of the breadboard with the minus side of the power rail – your circuits should now look something like this:

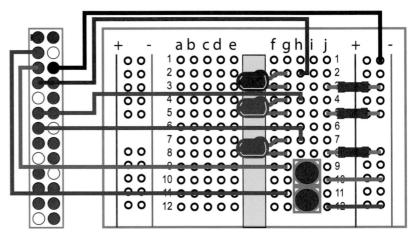

3 Turn on your Raspberry Pi then begin a new Python script by making attributes and methods available
import RPi.GPIO as GPIO
from time import sleep

4 Add statements to use board pin numbers in the script and set up 3 pins for output and 2 pins to listen for input
GPIO.setmode(GPIO.BOARD) ; GPIO.setup(13 , GPIO.OUT)
GPIO.setup(11 , GPIO.OUT) ; GPIO.setup(7 , GPIO.OUT)
GPIO.setup(5 , GPIO.IN) ; GPIO.setup(3 , GPIO.IN)

5 Then, add statements to initialize two variables with Boolean values to specify initial LED states
red_state = False
grn_state = False

6 Next, add a loop to illuminate the LEDs according to the variable values when no buttons are pressed

```
i = 0
while i < 8 :
        if ( GPIO.input( 3 ) and GPIO.input( 5 ) ) :
                GPIO.output( 7, red_state )
                GPIO.output( 11, False )
                GPIO.output( 13, grn_state )
```

The yellow LED is not controlled by either button so the loop ensures it remains turned off on each iteration.

7 Now, in the loop add statements to toggle a variable value when the first button gets pushed

```
        elif GPIO.input( 5 ) :
                red_state = not red_state
                i += 1 ; sleep( 0.2 )
                if red_state : print( str( i ) + ': RED ON' )
                else : print( str( i ) + ': RED OFF' )
```

8 Then, in the loop add statements to toggle a variable value when the second button gets pushed

```
        elif GPIO.input( 3 ) :
                grn_state = not grn_state
                i += 1 ; sleep( 0.2 )
                if grn_state : print( str( i )+': GREEN ON')
                else : print( str( i ) + ': GREEN OFF' )
```

The delay period in this example is reduced to two-tenths of a second so the response is rapid.

9 Finally, after the loop add a statement to clear the channels **GPIO.cleanup()**

10 Save the file, then enter this command to run the script and push the buttons to see the LEDs light up in response **sudo python gpio_inputs.py**

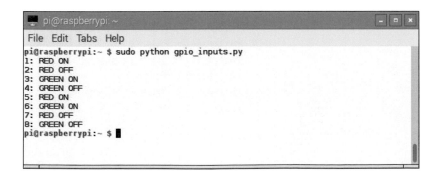

```
pi@raspberrypi:~ $ sudo python gpio_inputs.py
1: RED ON
2: RED OFF
3: GREEN ON
4: GREEN OFF
5: RED ON
6: GREEN ON
7: RED OFF
8: GREEN OFF
pi@raspberrypi:~ $
```

Controlling projects

The **while** loop controlling GPIO output and input can be made to run continually, simply by stating the test condition as **True**. Scripts can be halted by pressing the Ctrl + C key combination. This causes a KeyboardInterrupt exception that can be caught by including a try-except statement in the script. In order to clear the GPIO channels when a script exits in this way, you can import the Python **atexit** module and specify a function to be called on exit to its **register()** method:

gpio_control.py

If the GPIO channels are left uncleared, warnings will appear if the script is run again.

 Turn on your Raspberry Pi then begin a new Python script by making attributes and methods available
import RPi.GPIO as GPIO
from time import sleep
import atexit

 Add statements to use board pin numbers in the script and set up 3 pins for output and 2 pins to listen for input
GPIO.setmode(GPIO.BOARD) ; GPIO.setup(13, GPIO.OUT)
GPIO.setup(11 , GPIO.OUT) ; GPIO.setup(7 , GPIO.OUT)
GPIO.setup(5 , GPIO.IN) ; GPIO.setup(3 , GPIO.IN)

3 Then, define a function that will be called on exit to clear the GPIO channels when the script exits
def cleanup() :
    print('Goodbye.')
    GPIO.cleanup()
atexit.register(cleanup)

4 Next, add a statement to initialize a variable with a Boolean value and a function to run a sequence
run_state = False
def action() :
    if **run_state** :
        GPIO.output(7, True) ; sleep(0.2)
        GPIO.output(7, False)
        GPIO.output(11, True) ; sleep(0.2)
        GPIO.output(11, False)
        GPIO.output(13, True) ; sleep(0.2)
        GPIO.output(13, False)
    else :
        GPIO.output(7, False)
        GPIO.output(11, False)
        GPIO.output(13, False)

5 Add statements to display user instructions
```
print( 'Push Buttons To Start/Stop LED Sequence' )
print( 'Press Ctrl+C To Exit' )
```

6 Next, add a statement to catch the exception that occurs
when the user presses the Ctrl + C keys
```
try :
        # While loop to go here.

except KeyboardInterrupt :
        print( '\nScript Exited.' )
```

This example uses
the same circuits
configuration as the
previous example, as
illustrated on page 182.

7 Finally, insert a loop into the try clause to respond when
the user pushes either of the buttons
```
        while True :
                if ( GPIO.input(3) and GPIO.input(5) ) :
                        action()
                elif GPIO.input( 5 ) :
                        run_state = True
                        sleep( 0.2 )
                        print( 'Sequence Running...' )
                elif GPIO.input( 3 ) :
                        run_state = False
                        sleep( 0.2 )
                        print( ' Sequence Halted.' )
```

8 Save the file, then enter this command to run the script
and push the buttons to control the LED sequence
```
sudo python gpio_control.py
```

```
pi@raspberrypi: ~                                          _  □  ×
File  Edit  Tabs  Help
pi@raspberrypi:~ $ sudo python gpio_control.py
Push Buttons To Start/Stop LED Sequence
Press Ctrl+C To Exit
Sequence Running...
Sequence Halted.
Sequence Running...
Sequence Halted.
Sequence Running...
Sequence Halted.
Sequence Running...
Sequence Halted.
Sequence Running...
^C
Script Exited.
Goodbye.
pi@raspberrypi:~ $ ▮
```

Python's ability to
control input and
output via Raspberry
Pi's GPIO pins opens up
many possibilities when
connecting other devices,
such as electric motors.

Summary

- Raspberry Pi models have 26 or 40 GPIO (General Purpose Input Output) pins that can be controlled by Python scripts.

- GPIO pins can be addressed by their board pin number or by their BCM connection name on the Broadcom system chip.

- A breadboard lets you plug in and remove components for connection to the GPIO pins without soldering.

- Each LED (Light Emitting Diode) allows electricity to pass in only one direction so must be connected the right way round.

- A 270Ω (ohm) resistor can be added to a circuit to limit the current passing through an LED so it will not burn out.

- Methods and attributes to control the GPIO pins are provided in the Python **RPi.GPIO** module, and the **time** module provides a **sleep()** method that is useful to create delays.

- The **GPIO.setmode()** method specifies whether the script will use **BOARD** or **BCM** numbering to address pins.

- The **GPIO.setup()** method specifies the number of a pin to be used for output (**GPIO.OUT**) or for input (**GPIO.IN**).

- Power is supplied (**True**) or denied (**False**) to a specified pin number by the **GPIO.output()** method.

- The **GPIO.input()** method specifies a pin number to listen for input and can be used to recognize when a button gets pushed.

- Normally, the **GPIO.input()** method returns **True** but it returns **False** when a button is pushed so its condition can be tested.

- The **while** loop controlling GPIO input and output can be made to run continually by stating its test condition as **True**.

- Scripts can be halted by pressing Ctrl + C keyboard keys.

- The **GPIO.cleanup()** method clears GPIO channels and can be called automatically using the **atexit.register()** method.

Index

189

V

W

X

Z